Band 19 · Mai 1983 5

isotopenpraxis

AKADEMIE-VERLAG

ISSN 0021-1915 Isotopenpraxis, Berlin 19 (1983) 5, 145-184

EVP 9,- M

An unsere Autoren

Die Zeitschrift veröffentlicht Beiträge zu folgenden Themen: Kontrolle und Analyse radioaktiver und stabiler Isotope, Anwendung radioaktiver und stabiler Isotope in allen Gebieten der Forschung und Technik, der Industrie, Landwirtschaft und Medizin, Herstellung radioaktiver und stabiler Isotope sowie von markierten Verbindungen, kernelektronische Geräte für die Isotopenanwendung (Dickenmesser, Dichtemesser u. ä.) Detektoren für isotopentechnische Arbeiten und den Strahlenschutz in Isotopenlaboratorien, Meßtechnik und Strahlenschutz, Dosimetrie in Isotopenlaboratorien, kerntechnische Geräte für die Isotopenanwendung (Beton, Bestrahlungsanlagen u. ä.), Strahlenchemie, Radiochemie, ökonomische und perspektivische Fragen, Fragen der Weiterbildung.

1. Aufnahme von Manuskripten

Zur Veröffentlichung angenommen werden nur druckreife, noch nicht anderweitig publizierte Beiträge in deutscher, englischer oder russischer Sprache zum Themenbereich der Zeitschrift und zwar

- Originalarbeiten
- Kurze Mitteilungen
- Aus der Praxis
- Zusammenfassende Berichte
- Tagungsberichte
- Informationen.

Sind der Beitrag oder Teile daraus bereits veröffentlicht oder ist eine anderweitige Veröffentlichung vorgesehen, so muß dies der Redaktion mit genauer Angabe der Zweitveröffentlichung mitgeteilt werden. Mit der Annahme des Manuskriptes behält sich der Verlag das Recht der Vervielfältigung, Verbreitung und Übersetzung innerhalb der gesetzlichen Schutzfrist vor. Auszugsweiser Nachdruck oder photomechanische u. a. Wiedergabe sind an eine schriftliche Genehmigung des Verlages gebunden. Manuskripte, Zeichnungen und Photos von veröffentlichten Arbeiten werden nur auf den besonderen und bis zum Zeitpunkt der Veröffentlichung mitgeteilten Wunsch des Autors zurückgesandt.
Bei der unverlangten Einreichung von Manuskripten wird empfohlen, eine Befürwortung eines der Herausgeber oder Redaktionsausschußmitglieder beizufügen.
Manuskripte sind einzureichen bei der Redaktion, 1080 Berlin, Leipziger Str. 3–4

2. Form der Manuskripte

Die Autoren werden gebeten:
2 Manuskripte, eines davon in Original-Maschinenschrift, einseitig und mit 2zeiligem Abstand einzureichen; gut leserliche Photokopien sind zulässig; auf dem 1. Blatt (unter Weglassung der Bezeichnung)
den Titel der Arbeit
Vorname-Initialen und Namen sämtlicher Autoren
gegebenenfalls die Institution, in der sie tätig sind
eine kurze Zusammenfassung der Arbeit von höchstens 15 Zeilen
Keywords in alphabetischer Reihenfolge und
als Fußnote die Anschrift der Institution anzugeben;
wenn möglich, auf gesonderten Blättern Namen, Titel der Arbeit und Zusammenfassung in den beiden anderen obengenannten Sprachen 2fach beizufügen;
Literaturhinweise (fortlaufend numeriert), Fußnoten, Tabellen, Abbildungen (in 2–3facher Veröffentlichungsgröße), Abbildungsunterschriften (getrennt von den Abbildungen mit Zeichenerklärungen, Quellenangaben, falls erforderlich, und gegebenenfalls mit einem Vermerk für besondere Wünsche bezüglich der Abbildungsgröße), private und Dienstanschrift der Verfasser mit akademischen Titeln und ausgeschriebenen Vornamen auf besonderen Blättern ans Ende des Manuskriptes zu stellen;
Wünsche für den Kolumnentitel (Seitenüberschrift) bekanntzugeben;
Absätze durch Einzüge von mindestens 3 Anschlägen kenntlich zu machen; die Maßeinheiten des SI-Systems ("Système International") zu verwenden;
Korrekturen im Manuskript unmittelbar an der zu korrigierenden Stelle anzubringen;
Formeln deutlich lesbar und möglichst groß einzutragen;
Farbabbildungen und nachträgliche Änderungen oder Ergänzungen in Abbildungen zu vermeiden;
hervorzuhebende Textstellen zu unterstreichen, und zwar halbfetten Satz mit einer geraden schwarzen Linie, kursiven Satz (Schrägschrift) mit einer blauen Wellenlinie (alle Eigennamen). Sperrungen gestrichelt;
besondere Buchstaben durch farbige Unterstreichung zu kennzeichnen, und zwar Formelzeichen in
Griechisch-Kursiv (α, β, ...) rot,
Kursiv (A, B, ...) blau
halbfetter Kursiv ($\boldsymbol{A}$, $\boldsymbol{B}$, ...) 2 Linien blau (Vektoren),
halbfetter Grotesk (**A**, **B**, ...) 2 Linien violett (Tensoren);
bei Literaturangaben auf Vollständigkeit zu achten. In nachstehender Reihenfolge müssen angegeben sein:

a) bei *Zeitschriften*: Vorname-Initialen und Namen sämtlicher Autoren, Titel der Zeitschrift (nach „Periodica Chimica" oder möglichst wenig gekürzt), Bandnummer (unterstrichen), Jahreszahl (in Klammern), Heftnummer (nur bei Zeitschriften, die jedes Heft mit S. 1 beginnen) sowie die Seitenzahl, auf der die zitierte Arbeit beginnt.

b) Bei *Büchern*: Initialen der Vornamen und Name des Autors (Herausgeber werden durch den Zusatz „ed." gekennzeichnet), voll ausgeschriebener Titel des Buches in der Originalsprache, evtl. Bandnummer, voll ausgeschriebener Name des Verlages, Nummer der Auflage (außer der ersten), Erscheinungsort(e), Erscheinungsjahr (hier nicht in Klammern setzen!) und evtl. Kapitelnummer, Seitenzahl u. ä.

c) bei *Sammelbänden, Tagungsberichten*: Es soll zusätzlich der Titel des zitierten Beitrags angegeben werden. Die Seitenzahl muß unbedingt angegeben werden. Aus dem Zitat muß hervorgehen, ob es sich um einen Tagungssammelband oder um einen noch nicht veröffentlichten Vortrag handelt. Zwischen Tagungsort und Erscheinungsort des entsprechenden Tagungsberichts ist zu unterscheiden.

3. Korrekturen

Der Autor erhält zwei Umbruchabzüge, von denen einer möglichst schnell korrigiert an die Redaktion zurückgereicht werden soll. Auf Wunsch des Autors kann die Korrektur auch von der Redaktion gelesen werden.
Durch nachträgliche Manuskriptänderungen verursachte Korrekturkosten gehen, sofern sie 10% der Satzkosten übersteigen, zu Lasten des Autors.
Der Autor erhält 50 Sonderdrucke kostenlos. Weitere Sonderdrucke können bis zur Rückgabe der Korrekturabzüge auf Kosten des Autors bei der Redaktion bestellt werden.

Bezugsmöglichkeiten

Bestellungen sind zu richten
– **in der DDR** an den Postzeitungsvertrieb unter Angabe der Kundennummer des Bestellenden oder an den AKADEMIE-VERLAG, DDR-1086 Berlin, Leipziger Straße 3–4, PF-Nr. 1233
– **im sozialistischen Ausland** an eine Buchhandlung für fremdsprachige Literatur oder an den zuständigen Postzeitungsvertrieb
– **in der BRD und Berlin(West)** an eine Buchhandlung oder an die Auslieferungsstelle KUNST UND WISSEN, Erich Bieber OHG, D-7000 Stuttgart 1, Wilhelmstraße 4–6
– **in den übrigen westeuropäischen Ländern** an eine Buchhandlung oder an die Auslieferungsstelle KUNST UND WISSEN, Erich Bieber GmbH, CH-8008 Zürich, Dufourstraße 51
– **im übrigen Ausland** an den Internationalen Buch- und Zeitschriftenhandel; den Buchexport, Volkseigener Außenhandelsbetrieb der Deutschen Demokratischen Republik, DDR-7010 Leipzig, Postfach 160; oder an den AKADEMIE-VERLAG, DDR-1086 Berlin, Leipziger Straße 3–4

Zeitschrift „Isotopenpraxis"

Herausgeber: Im Auftrag des Akademie-Verlages herausgegeben von Prof. Dr. Justus Mühlenpfordt
Verlag: Akademie-Verlag, DDR-1086 Berlin, Leipziger Straße 3–4, PF-Nr. 1233; Fernruf: 2236229; Telex-Nr.: 114420;
Bank: Staatsbank der DDR, Berlin, Kto.-Nr.: 6836-26-20712
Verantwortlich für den Inhalt des Heftes: Prof. Dr. H. Hübner, Leipzig
Anschrift der Redaktion (Fernruf Frau H. Schier: 6351177) der Zeitschrift Isotopenpraxis: Akademie-Verlag, DDR-1086 Berlin, Leipziger Str. 3–4, PF-Nr. 1233
Veröffentlicht unter der Lizenznummer 1527 des Presseamtes beim Vorsitzenden des Ministerrates der Deutschen Demokratischen Republik
Gesamtherstellung: VEB Druckerei „Thomas Müntzer", DDR-5820 Bad Langensalza
Erscheinungsweise: Die Zeitschrift „Isotopenpraxis" erscheint monatlich. Die 12 Hefte eines Jahrganges bilden einen Band. Bezugspreis je Band 180,– M zuzüglich Versandspesen Preis für die DDR 108,– M) Preis je Heft 15,– (Preis für die DDR 9,– M)
Bestellnummer dieses Heftes: 1079/19/5

AN (EDV) 12831

ISOTOPENPRAXIS

Band 19 (1983) Heft 5, S. 145–184

Redaktion und Verlag: Akademie-Verlag, Redaktion Isotopenpraxis, 1086 Berlin, Leipziger Straße 3–4

Inhaltsverzeichnis

Contents

P. Popp, J. Leonhardt

Charakterisierung von coulometrischer und hypercoulometrischer ECD-Betriebsweise (*in Englisch*)

Isotopenpraxis **19** (1983) H. 5, S. 160 bis 162
4 Abb., 1 Tab., 10 Lit.

Die coulometrische Betriebsweise liegt vor, wenn jedes Probemolekül das in den ECD gelangt, in ein negatives Ion umgewandelt wird und jedes negative Ion zur Absenkung des Ionisationsstromes beiträgt. Die Empfindlichkeit des Detektors beträgt in diesem Fall 96500 C/mol. – Eine hypercoulometrische Betriebsweise ist dann gegeben, wenn Bruchstücke aus dissoziativen Anlagerungsprozessen im Detektor mehrmals an der Bildung negativer Ionen teilnehmen und verstärkende Wirkungen von Raumladungen genutzt werden. In diesem Fall überschreitet die Empfindlichkeit des ECD den durch die *Faraday*sche Zahl gegebenen oberen Grenzwert. – Der Aufbau beider Detektoren wird beschrieben.

M. M. Abou Sekkina, N. Hamami, S. A. Abou El-Enein

^{60}Co-Gammastrahlungsschäden an polykristallinen Acetylcholinhalogenid-Elektrolyten (*in Englisch*)

Isotopenpraxis **19** (1983) H. 5, S. 163 bis 165
4 Abb., 3 Tab., 7 Lit.

Von der Fa. BDH gelieferte feste Proben von Acetylcholinhalogenid $[CH_3COOCH_2CH_2 \cdot N(CH_3)_3]X$, bei denen X = Br, I, wurden gereinigt und aus einem geeigneten Lösungsmittel umkristallisiert. Umfangreiche Untersuchungen der IR-Absorptionsspektren sowie Analysen der Röntgenbeugungsdiagramme wurden sorgfältig durchgeführt, und zwar vor und nach der Einwirkung energiereicher Gammadosen ($5 \cdot 10^6$ rd). Die erzielten Ergebnisse wiesen darauf hin, daß Azetylcholinbromid eine höhere Strahlenbeständigkeit hat als Azetylcholinjodid. Diese Tatsache wurde ausgehend von den Unterschieden in der Bindungsenergie und in der elektronegativen Ladung diskutiert. Sie könnten auch anhand der durch die absorbierte Gammadosis hervorgerufenen Schäden und der sie begleitenden Bildung von Gitterstörstellen, kombiniert mit Donatorstörstellenzentren, die bei Jodidsalzen sehr ausgeprägt auftreten, diskutiert werden. Damit konnten die IR-Absorptionsspektren und die Röntgenbeugungsanalyse wirksam als schnelle und empfindliche Mittel zur Strahlungsdetektion unter Verwendung von Azetylcholinhalogenid-Polykristallen als Detektoren eingesetzt werden.

Kurze Originalmitteilungen

L. Buchmann, T. Grätsch, K. Wagner

Charakterisierung von Reinstgraphit durch Neutronenaktivierungsanalyse und Autoradiographie

Isotopenpraxis **19** (1983) H. 5, S. 166 bis 168
5 Abb., 2 Tab., 6 Lit.

J. Römer, S. Schwarz, G. Weber

Synthese des 3-Pyrrolidinosulfonylesters von 17α-Ethinyl-[9α, 11α-^{3}H]-estradiol

Isotopenpraxis **19** (1983) H. 5, S. 168 bis 170
2 Abb., 1 Tab., 5 Lit.

Tagungsberichte

Isotopenpraxis **19** (1983) H. 5, S. 170

Informationen

Isotopenpraxis **19** (1983) H. 5, S. 171 bis 177

Buchbesprechungen

Isotopenpraxis **19** (1983) H. 5, S. 178 bis 184

P. Popp, J. Leonhardt

Characterization of Coulometric and Hypercoulometric ECD Mode of Operation (*in English*)

Isotopenpraxis **19** (1983) No. 5, p. 160 to 162
4 Fig., 1 Tab., 10 Ref.

The coulometric mode of operation is given, if each sample molecule passing the ECD is converted into a negative ion and if each negative ion causes a decrease of the ionization current. The sensitivity of the detector in this case amounts to 96500 C/mol. A hypercoulometric response is given, if fragments from dissociative processes undergo electron capture at serveral times and if the enhancing effect of space charges is used. In this case the sensitivity of the ECD exceeds the upper limit given by *Faraday's* number. The operation conditions of both ECD-modes are described.

M. M. Abou Sekkina, N. Hamami, S. A. Abou El-Enein

^{60}Co Gamma Radiation of Damage of Polycrystalline Acetyl Choline Halide Electrolytes (*in English*)

Isotopenpraxis **19** (1983) No. 5, p. 163 to 165
4 Fig., 3 Tab., 7 Ref.

Solid samples of acetylcholine halide $[CH_3COOCH_2CH_2 \cdot N(CH_3)_3]X$, where X = Br, I as supplied from (BDH) company have been purified and recrystallized from the appropriate solvent. Extensive studies of infrared absorption spectra and analysis of the X-ray diffraction patterns were carefully conducted before and after exposure to high energy gamma absorbed dose (5×10^6 rd). Results obtained indicated that acetylcholine bromide has higher radiation resistance than acetylcholine iodide and this was discussed on the basis of bond strength and electronegativity differences. These could be also discussed in terms of gamma absorbed dose damage and their attendent creation of lattice imperfections in combination with donor defect centres which are liable to occur pronouncedly for the iodide salt. Finally, the IR absorption spectra as well as X-ray diffraction analysis could find their great usefull application as a rapid and sensitive tools for radiation detection at high efficiency using acetylcholine halides polycrystals as detectors.

Letters to the Editor

L. Buchmann, T. Graetsch, K. Wagner

Characterization of Extra-Pure Graphite through Neutron Activation Analysis and Autoradiography

Isotopenpraxis **19** (1983) No. 5, p. 166 to 168
5 Fig., 2 Tab., 6 Ref.

J. Roemer, S. Schwarz, G. Weber

Synthesis of the 3 Pyrrolidinosulphoxylic Ester of 17α-Ethinyl-[9α, 11α-^{3}H]-Estradiol

Isotopenpraxis **19** (1983) No. 5, p. 168 to 170
2 Fig., 1 Tab., 5 Ref.

Meetings

Isotopenpraxis **19** (1983) No. 5, p. 170

Information

Isotopenpraxis **19** (1983) No. 5, p. 171 to 177

New Books

Isotopenpraxis **19** (1983) No. 5, p. 178 to 184

Isotopenpraxis 19 (1983) 5, S. 145–153

Deuterium and ^{18}O Variations in Lakes of the Schirmacher Oasis (East Antarctica)

W. Richter, G. Strauch

(Zentralinstitut für Isotopen- und Strahlenforschung der AdW der DDR, Leipzig)[1])

As a snow and icefree rock area the Schirmacher Oasis (Dronning Maud land) is situated on the coast of the Antarctic continent, between inland and shelf ice. In the area of the oasis different exogenic conditions have produced multiform types of lake basins.
In the present report first results about isotope hydrological relations in the lakes of the Schirmacher Oasis are discussed. The lakes can be classified by isotope hydrological parameters into different groups. The classification depends on in- and outflow, evaporation processes, nature of inflow and location.

Keywords
antarctica; deuterium; hydrology; isotope ratio; lakes; oxygen 18; stable isotopes

1. The geography of the Schirmacher Oasis

Even the first explorers who went to Antarctica reported icefree areas on the forbidding continent. Not until the time of *Rymill* and *Stephenson* [1], however, were parts of the Antarctic Peninsula which were not covered by snow and glaciers compared to oases. The Schirmacher Lake District was discovered during the German Antarctic Expedition of 1938/39 [2], and the Bunger Oasis as a result of American exploration from the air. The original term Schirmacher Lake District was logically assigned to the category of Antarctic oases only a few years after *Ritscher's* publication of 1942 [3]. The initial idea of the oasis was limited to ice-free parts of the Antarctic coast [4]. This was later extended to include snow and icefree mountain regions (e.g. in the Wohlthat Massiv, Dronning Maud Land). In East Antarctica alone there are, between 10°E and 130°E, about 12 major oases, normally near the coast. The Schirmacher Oasis has a small area and belongs to the group of typical oases which can also be described as close to the shelf [5]. Its situation is given by the coordinates of astronomically surveyed observation points East (Novolazarevskaya) and West, as follows [6, 7]:

East	−70°46′04″ ± 13″	−11°49′54″ ± 48″
West	−70°44′21″ ± 04″	−11°26′03″ ± 02″

As far as size is concerned, the following data can be given (in km², acc. to ATLAS ANTARKTIKI [5]):

Region	Area, general	in a narrower sense	Bed-rock	La-goons	Firn/ice fields	Lakes 6
1,000	35	31	27	3	3	2

The heigth of the oasis is between 0 and 228 m above the level of the epishelf lakes. The average heigth of the rolling relief of the oasis with its dome-shaped hills is around 100 m. There are two nuntaks 2 km to the south, and an island is situated to the north (Fig. 1). Selected climatological parameters (1961–66 series) [5]

Average air temperatures, °C			Wind	Direc-tion	Precipi-tation
Annual	July	January	m/s		mm/a
−11·0	−18·2	−1·5	10·2	ESE	309

The amount of precipitation does not reflect the quantity that is actually available in the water balance of the oasis. Snow accumulates only locally on the lee side of obstacles,

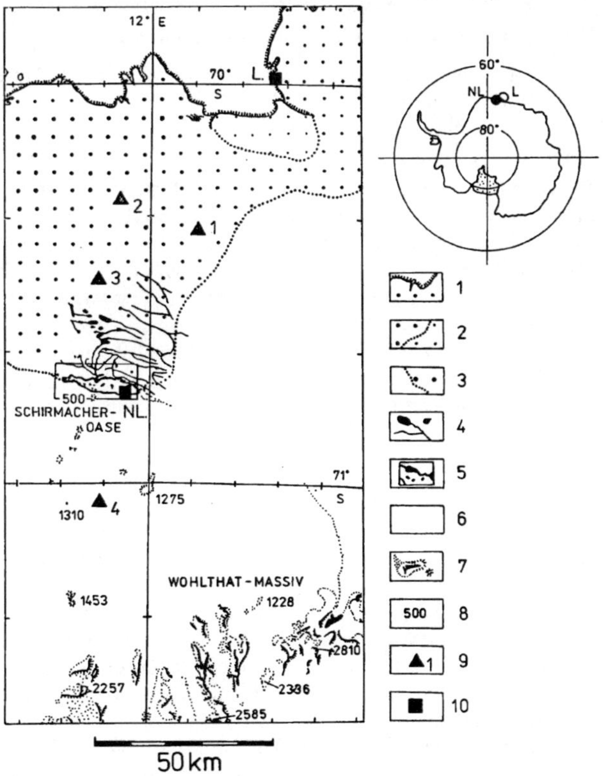

Fig. 1. Dronning Maud land, region Schirmacher Oasis
1 – shelf ice border; 2 – shelf glacier; 3 – border shelf ice – inland ice; 4 – rivers, lakes; 5 – Schirmacher Oasis; 6 – inland ice; 7 – ice free areas; 8 – altitude above sea level, in metres; 9 – ice drilling holes; 10 – Novalazarevskaya NL, Lazarevskaja L.

[1]) Address: DDR-7050 Leipzig, Permoserstr. 15

while most of the oasis remains snow-free because of the wind.

2. The lakes of the Schirmacher Oasis

Lakes are an important though not necessarily dominant geographic feature of the snow and ice-free rocky landscapes that are the oases of the East Antarctic. The first name given to the Schirmacher Oasis [2] indicates that they constitute a major landmark in the area. Lakes in this connection means all bodies of water in naturally created glacial throughs or basis bordered by dams on the land surface. The definition given by *Forel* [8] which continues to be widely accepted today, expressly excludes a direct connection with the sea (this is also said by such authors as *Hendl*, *Jäger*, and *Marcinek* [9]). This definition, however, makes no allowance for epishelf lakes in Antarctic cases (the first indirect reference to this type is made by *Korotkevich* [10], ATLAS ANTARKTIKI [5], *Simonov* [4], *Korotkevich* [11]). These are regarded as a subtype of the lagoon lakes. At the foot of the Schirmacher Oasis they are separated from the open sea by the enormous ice sheet of the shelf glacier (Fig. 1) but are directly connected to the sea by a passage under the shelf ice. The hydrochemistry, thermal conditions and hydrobiology of these lakes make them comparable to fresh-water lakes on the mainland. Their special characteristics include the situation on the sea shelf and the regular occurrence of tides. The level to which they are filled with fresh water depends on the thickness of the meltwater stratum retained by the shelf glacier and is a few meters above sea level.

2.1. The genesis of the lake basins in the Schirmacher Oasis

While on the one hand distinguishing between

1. snow and ice-free rock surface in the oasis and the Aerodromnaya nunatak,
2. the inland glacier, and
3. the shelf glacier

one can at the same time attribute the formation of the lake basins to different exogenous factors. For the area of the Schirmacher Oasis these may be subdivided as described in the following, with reference to the ATLAS ANTARKTIKI II [5]. Where a comparison is possible with *Hutchinson* [12] (whose classification No. is given in brackets) this is indicated. In some cases differentiation goes beyond these references, as a result of the scope of field work carried out and in response to the requirements of the present paper.

2.1.1. Glacial rock basins

2.1.1.1. Tectonic- exarative genesis (type 26, Hutchinson [12]). This type of lake may emerge in fault and shatter zones as a result of glacial action on zones of weakness in solid rock. In the Schirmacher Oasis no clear demarcation is possible in most cases from the following type. This would require detailed tectonic, and the cover of congelifracts and moraine debris on the surface of the oasis interferes with field investigations.

It is highly probable that Lake Dlinnoye belongs in this category because of its length, the strike direction of its longitudinal axis, and the geomorphology and hydrography of the surrounding area. It is uncertain whether tectonic factors were involved in forming the basins of Lake Zapadnoye and Lake Podprudnoye. The latter is situated southwest of Lake Dlinnoye in the strike direction of the valley-type landform partly filled by that lake.

2.1.1.2. Lake basins of exarative origin. These presumably constitute the majority of the water-filled depressions in the Schirmacher Oasis. Their size varies considerably, and many of these lakes are quite shallow (with depths ranging from a few meters to several decimeters). Largest in this group is Lake Zub with an area of 0·5 km² (max. depth 5 m, average 2 m). With the exception of a few lakes, most of them larger in size, we have no data on areas and depths in the Schirmacher Oasis. In some cases the shorelines have changed with the aerial photogrammetric survey of the region conducted in 1961, due to the thawing of snow, firn and ice fields (lake 129 m, ca. 1 km south of Mt. Primetnaya). In other cases the draining of ice wall dammed lakes has led to a lowering of the water level and changes in shoreline and depth, as happened with Lake Podprudnoye in March 1979. Lake Prilednikovoye and Lake Iskristoye are possibly of the glacier-dammed type and may disappear more or less completely as the inland glacier recedes. In small catchment areas in the oasis where there is no outlet, small and minute evaporation lakes may arise.

2.1.1.3. Morainic lakes (type 21). Solifluction and the uniform character of the glacier front to the south of the Schirmacher Oasis have prevented the development of moraine-dammed lakes in the area of the inland glacier. However, on the southern edge of Lake Ozhidaniya evidence has been found, in the rocky precipice of the oasis, of a 200 – 300 m long moraine dam. Its formation probably correlates with an advance of the shelf glacier which formerly extended to that point and which occurred in the distant past. The moraine dammed a valley-type through directed to the northeast and about 1·5 km long. Two lakes are left of what was once a larger moraine-dammed lake. Where the dam joins the precipice in the southeast, it is severly eroded and the level of the earlier lake has been drastically lowered.

2.1.1.4. Glacier-dammed lakes (type 24). An important characteristic of this type is the formation of a lake basin when a glacier acts as a dam. In the case of the Schirmacher Oasis it is problematic to assign individual lakes to this type in view of the insufficient knowledge we have about the subglacial relief. A distinct terrace above the shore of Lake Prilednikovoye suggests a higher water level at some time in the past as a result of glacier damming. On the other hand, Lake Iskristoye discharges at the highest possible level.

2.1.1.5. Icewall-dammed lakes. In the central part of the Schirmacher Oasis enormous snowdrifts turned into ice form several walls which block the valley between Lake Podprudnoye and Lake Dlinnoye. These ice formations are of the type of the "glacier formed by drifting snow" and are comparable to stationary dense firn ice. The icewalls act as dams for several lakes and have raised the level of Lake Podprudnoye by 3 – 5 m. This type of lake may exist for several years and develops its own biological and thermal conditions. In the Schirmacher Oasis it normally takes the form of small or very small lakes with a depth of a few meters and an area of between 10 and several thousand m², depending on the situation and genesis. In summers with a favourable climate, e.g. 1978/79, the icewalls may give way and subglacial channels ("ice caverns") may form. (However, these drainage channels also exist in firn ice fields, particularly in the eastern part of the Oasis.) As a result

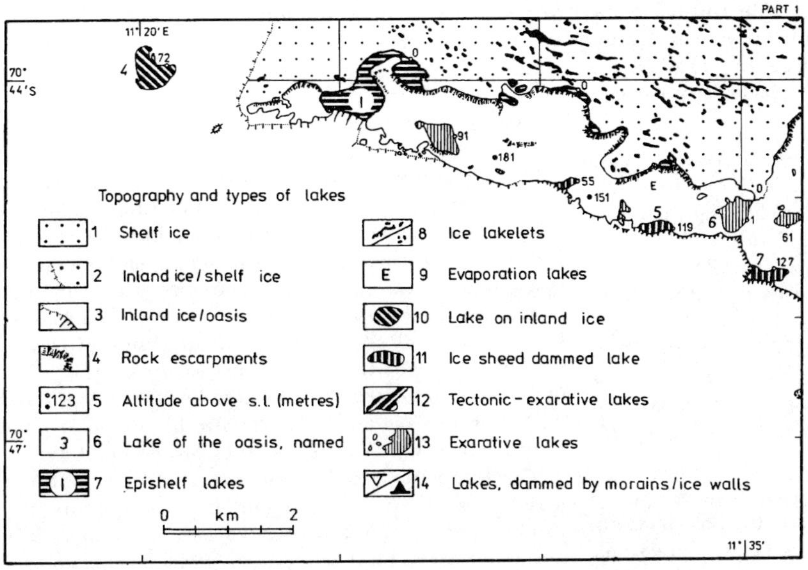

Fig. 2a. Schirmacher Oasis. Topography and types of lakes. I

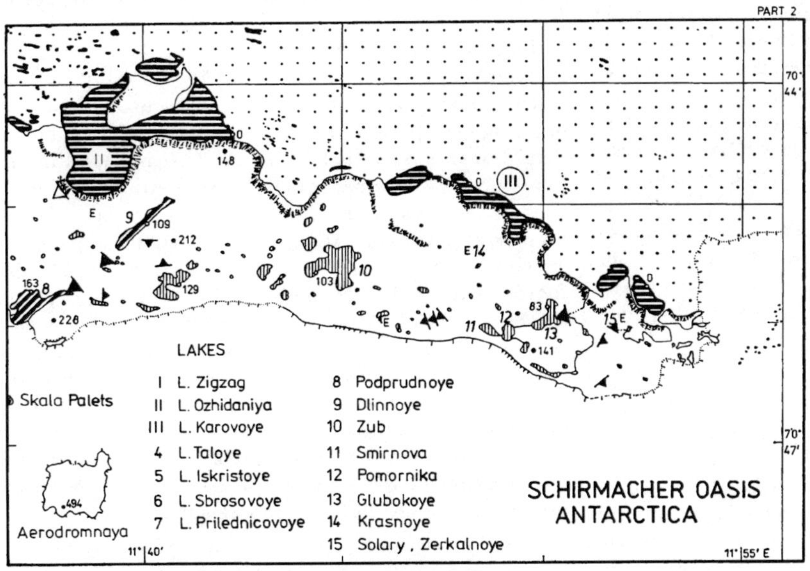

Fig. 2b. Schirmacher Oasis. Topography and types of lakes. II

of such ruptures the meltwater lakes normally disappear, or the direction of discharge changes after reductions in water level as happened in the case of Lake Podprudnoye. Lakes of this type observed during the 1979/80 mapping period appear separately in Fig. 2.

2.1.2. Lake basins in the area of the inland galcier (type 23)

The inland glacier forms the southern boundary of the Schirmacher Oasis. Its surface appears more dome-shaped in the eastern part, and the rest is stepped in the strike direction with long undulations. The curvature is mostly convex. Between the oasis and the Aerodromnaya nuntak the slope has an extended concave shape over a maximum length of 2 km. In the extreme west of the oasis a hollow is formed in connection with ice damming whose strike direction is northeast and later eastnortheast. At its northern edge is the ice trough of Lake Taloye, a comparatively large and shallow lake on the inland glacier. Its ice cover partly rests on the ground, possibly as a result of recent reductions in level. The longest axis is almost 750 m.

No other remarkable lakes were found in the area of the inland glacier. The reason must be sought primarily in the curvature of the foot and lower slope portions of the glacier. Cryoconite holes and similar depressions on a miniature scale (type 23a in *Hutchinson* [12]) with diameters of a meter or less form at heights up to between 350 and 450 m. Their size, number and water level differ from one year to the next, and *Hutchinson* [12] remarks that they are interesting mainly as "limnological curiosities". Almost none were formed in the summer of 1979/80, in contrast with the previous year.

2.1.3. Lake basins in the area of the shelf glacier

The northern edge of the Schirmacher Oasis consists almost exclusively of sea shore, and the shelf glacier farther to the north rises from several meters to more than 40 m. In the western half of the shelf glacier a block field extends near the oasis which is more than 3 km wide and has complicated hydrographic conditions. It includes many small and minute lakelets, and there are large and shallow ice lakes farther away on the undisturbed glacier surface but outside the area studied.

2.1.3.1. The basins of the epishelf lakes. Meltwater lakes appear for the first time on plate 50 of *Ritscher's* paper [2], at the foot of the exarated north wall of the Schirmacher Oasis, in its eastern part. *Korotkevich* [10] found that these lakes were connected to the open sea (tidal walls and fractures). Compared with the other lakes in the oasis these are large to very large and normally deep (Lake Ozhidaniya 1·4 km², max. depth 105 m; Lake Zigzag 122 m deep, sounded), with a permanent ice cover. These lakes collect the meltwater from the shelf glacier roof slope, the direct discharge from the inland glacier and water from the Schirmacher Oasis.

2.1.3.2. Lake basins on the shelf glacier. The shape of these basins is directly related to the structure of the shelf glacier. Where the surface is undisturbed, large depressions may form which are comparable to Lake Taloye or possibly even much more extensive (Fig. 1, ca. 15 km north of the oasis). There are no such lakes in the oasis proper, and here they mostly take the form of narrow, small and often groove-type bodies in the block field of the shelf glacier. In Fig. 2 these lakes designate the strike direction of ice blocks outside the western part of the oasis in a typical manner. So far the limnology of these lakes between the 25–35 m high blocks has not been studied, nor do we have descriptions of the larger lakes farther north on the undisturbed shelf glacier. In summer the area under study is barely accessible on foot, and there are hardly any routes for tracked vehicles because of the high relief energy of the block field. It seems that at least the smaller lakes freeze down to the floor in winter, and their ice cover often rises up to a height of much more than 1 m.

2.1.4. Feed conditions, ice cover and thermal conditions

The feed conditions in the Schirmacher Oasis depend primarily on the meltwater supply from the inland glacier (oasis plateau) and from the shelf glacier (foot of the oasis). Of lesser importance are the snow and ice fields in the oasis itself. Most of the lakes in the Schirmacher Oasis have the typical characteristics of open lakes, and there are only few examples of inland catchment with areas practically no effluent (Fig. 2). All the effluent is collected in the epishelf lakes and goes to the sea through tunnels under the glacier. The discharge of meltwater starts sometime in November and is largely finished by the end of February. The effluent quantities vary with meteorological conditions, and no studies have been made so far.

It is only logical that differences should occur in the formation of the ice covers, considering the diversity of lake basins, their different areas and depths, the floors which may be rock or ice, the situation with regard to glaciers and the origin of melt or feed water. The results are shown in table 1. Permanent ice covers (2–3 m thick) are characteristic of lakes which border on glaciers and are sufficiently deep (more than 5 m), for example Lake Iskristoye. They also occur on sufficiently large lakes on the glacier (Lake Taloye) and on the largest and deepest epishelf lakes. In the case of the icewall-dammed lakes the crystal structure also suggests a trend toward permanent ice covers. The crystals have few bubbles and are polygonal-prismatic with diameters of approximately 5 cm or more and entrapped air in the form of bubbles or clusters on the crystal faces. The surface of the ice cover is serrated as a result of evaporation, and from the underside at least some of the long crystal points extend into the free water. It can be assumed that these ice covers open only at large and possibly secular intervals.

The transition to the perennial type of ice cover is difficult to establish, but the icewall-dammed lakes and the chain consisting of Lake Smirnova, Lake Pomornika and Lake Glubokoye problably belong in this category. This type tends toward lakes with temporary ice covers in view of the ice thickness and crystal structure and considering the structure of the surface and underside of the ice. The intermediate position is due primarily to the alternation of warm summers with intense radiation and cool summers with low radiation. Under favourable climatic conditions the ice cover on these lakes will open completely.

The temporary ice covers with small prismatic crystals which measure only a few cm² in cross section grow to a thickness of about 2 m each winter. They have a smooth underside and a similar top side (which is rippled and wavy after the end of the winter). Shallow lakes lose this ice cover completely in a "normal" polar summer, but in a cold year only small and minute shallow lakes can be expected to be ice-free.

The thermal characteristics of the lakes in the Schirmacher Oasis make them belong to the stratification type of the polar lake, "with surface temperatures always below 4 °C, ice-free period very short, and circulation only at the height of summer" [12]. When the thermal conditions in the lakes of the Schirmacher Oasis were studied it was found that the situation of a lake in a region with polar climate alone is not sufficient to prevent the water temperature from exceeding the value for the maximum density. Important factors involved here include the area, depth, feed and discharge, wind exposure, ice thickness and the situation of a lake near or on a glacier. Apart from glacial lakes with largely thermostatic conditions at temperature of only a few tenths of a degree Centigrade above zero, there are all kinds of transitional stages ending with pond-type bodies which may heat up considerably. This means that the temperature factor is merely a characteristic feature of the larger polar lakes but not an exclusive factor.

The temperature profiles of the epishelf lakes which do not become ice-free have been well illustrated by *Simonov* in Fig. 29 [4]. In this type of lake the temperatures rise from 0 °C under the ice cover to about 0·2 °C and remain constant in the thermally unstructured hypolimnion (Fig. 7). The larger lakes and particularly those of sufficient depth in the oasis best comply with the mixing and stratification type of the polar lake. There are, however, in this area which has been ice-free for only a few thousand years, many small and minute overdeepened depressions filled with water to levels ranging from less than a meter to several meters. They exist side by side with large shallow lakes, normally freeze down to the floor in winter and many heat up considerably under favourable conditions. *Simonov* [4] reports water temperatures between 10 and 17 °C for Lake Solary in the eastern part of the Schirmacher Oasis. We were able to confirm this for Lake Zerkalnoye which is in the immediate vicinity. Where feed is insufficient these ponds not only heat up extensively but also lose considerable amounts of water through evaporarion [13]. This thermal type modifies the definition of the polar lake. The most striking example of a lake with high evaporation is Lake Krasnoye. Two distinct terraces point to a gradual but drastic lowering of the water level, but the hydrochemical and isotope values are not comparable to the phenomenon of the Antarctic salt lakes such as Lake Vanda and Lake Bonney [14]. The Schirmacher Oasis does not have the necessary conditions for this.

2.2. *Deuterium and ^{18}O variations in the lakes*

2.2.1. *Methods*

The Antarctic summer months were chosen for sampling with the aim of investigating the dynamics of the lakes in question. Samples were taken from mid-November to mid-February, the frequencies being once a week for shallow exaration basins, once every two to four weeks for larger lakes and only once for epishelf lakes. Where sampling was repeated the location on the lake was determined with the aid of a compass. A 1-liter Ruttner sampler with built-in thermometer was used, and the sampling points were 1 m apart for medium deep and deep lakes and 0·5 m apart in the case of Lake Zub. From shallow lakes samples were taken on the surface near the shore. Where there was an ice cover we employed a sampling winch, while a rubber dinghy was used in the absence of ice. A valve sampler was employed in the summer of 1979/80.

Since the Schirmacher Oasis is separated from the sea by a shelf belt which is between 80 and 100 km wide, the hydrochemistry of its lakes is much different compared with the genuine coastal oases (Ongul Island, Victoria Land, Vestfold Oasis). For example, the maritime influence is limited to atmospheric aerosol input, weathering is mostly of the physical type and weathering products are introduced from the mountain range in the south of Dronning Maud land (with a high potassium content from feldspars.) All this is reflected in the hydrochemical properties of the meltwater and lakes.

The Soviet Antarctic Expedition of 1965/66 conducted hydrochemical studies of the lakes in the oasis. Zoning was possible in the eastern part of the oasis according to the degree of mineralization and the salt composition [15]. The first group (mineralization degree 1 mg equivalent per liter and Na/K greater than 2) comprised all shelf lakes and the small exaration lakes close to the northern edge of the oasis (Rozovoye, Krasnoye). Although the epishelf lakes are connected with the sea and their hydrochemistry is the same as for seawater, mineralization is 350 to 400 times less and there is a high potassium content which stems from the oasis. In the case of the other exaration lakes the degree of mineralization is linked to the evaporation process. These lakes form the second group (including subgroups) where mineralization is less than a mg equivalent per liter and Na/K less than 1. In the direction of the inland glacier the sulfate and chloride contents drop in favour of hydrogen carbonate (the same can be said for the magnesium content), and this is attributed to the growing "continental influence" by *Simonov* [4].

Lake Glubokoye is a large meltwater reservoir and therefore has a high K content (60 percent of the cations). A particular characteristic of Lake Zub is the prevalence of alkali chlorides (63 per cent), and this can be attributed to the extensive surface (0·460 km^2) and maritime precipitation [16, 17].

Compared with the saline lakes of the East Antarctic coastal oases all stagnant waters of the Schirmacher Oasis can be described as freshwater lakes.

All samples were kept in airtight polyethylene or glass bottles and sealed. The samples were analyzed at ZfI in Leipzig and converted into hydrogen measuring gas for δD determination acc. to the Chromium method [18].

Three double tests were carried out on each sample. The reference gas was a laboratory standard LLWS,

$\delta D = -63{\cdot}5$ ‰ V-SMOW and a working standard (Antarctic precipitation, D = -229 ‰ V-SMOW). The deuterium concentration in the usual δ-scale is related to Vienna-SMOW (IAEA, Vienna).

$$\delta = \left[\frac{\left(\frac{R'}{R}\right)_{\text{Sample}}}{\left(\frac{R'}{R}\right)_{\text{Standard}}} - 1 \right] \times 10^3 \text{ ‰}$$

where R'/R is the ratio of the heavy to the light nuclide in at.%. The determination of the ^{18}O abundance is similar. The measuring gas is CO_2 which is brought into an equilibrium of exchange with the water samples in question [18].

Tab. 1. Ice conditions on the lakes of the Schirmacher Oasis (a survey)

1.	Permanent ice covers	
1.1.	Schirmacher Oasis	Lake Iskristoye
		Lake Smirnova (?)
1.2.	Inland glacier	Lake Taloye
1.3.	Shelf glacier	Lake Zigzag
		Lake Ozhidabiya
		Lake Predgornoye (?)
		Lake Karovoye (?)
		Lake Privalnoye (?)
		Lake Kholodnoye (?)
2.	Permanent-temporary ice covers	
2.1.	Schirmacher Oasis	Lake Smirnova
		Lake Pomornika
		Lake Glubokoye
		Lakes 129 at Mt. Primetnaya
2.2.	Shelf glacier	only shallow meltwater basins
3.	Temporary ice covers	all other lakes in the ice-free area of the oasis

2.2.2. *Results and discussion*

Several papers have been written about the physical and chemical analysis of Antarctic salt lakes with ice covers [19–22]. The isotopic ratios for deuterium and ^{18}O enabled us to draw important conclusions with regard to the genesis, age and condition of the Antarctic saline lakes and their water balance.

By contrast, few studies have been made of the normal type of Antarctic freshwater lake with its extremely low salt content. An attempt was therefore made in connection with the GDR research program related to environmental isotopes in the Schirmacher Oasis, to select characteristic types from the multitude of lakes for isotope hydrology sampling and investigation. First results have led to the following classification:

- small and shallow exaration lakes (ponds)
- large and shallow exaration lakes
- exaration lakes of medium depth (ca. 10 m)
- deep exaration lakes
- epishelf lakes
- lakes on glacial ice

2.2.2.1. Deuterium and ^{18}O variations in small and shallow exaration lakes. The inland bodies of water of this type are small and shallow ponds which are seldom more than 1 m deep and up to 100 m wide. They are normally the first to thaw and become completely icefree. Under conditions of favourable radiation and low winds they can reach temperatures of more than 15 °C. The water temperature fluctuates heavily as a function of the prevailing wind and clouds (Fig. 3). The isotopic ratios depend on the water balance, and three subtypes can be distinguished.

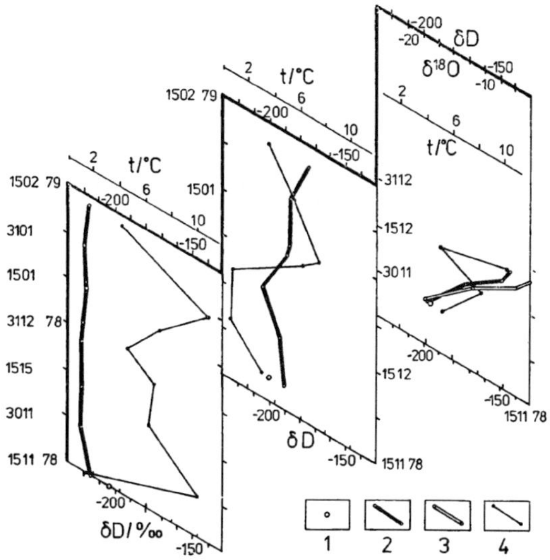

Fig. 3. Three types of smallest and shallow exaration lakes: Lake Zenit, Lake Krasnoye and Lake Zerkalnoye (from left to right). δD and δ^{18}O-variations

1 — δD ice cover; 2 — δD water; 3 — δ^{18}O water; 4 — temperature

Subtype No. 1 — Lake Zenit: There is a balance between the meltwater feed from snow and ice fields and from the inland glacier or other lakes, and the discharge. The delta values for D and ^{18}O are the same as for the input. Lake Zenit receives meltwater from snowdrifts and firn ice (δD around -220 ‰).

Subtype No. 2 — Lake Krasnoye: Here the feed is less than in the case of subtype No. 1, and no discharge takes place. Considerable quantities of water are lost by evaporation over longer periods. The δD value drops until the lake becomes ice-free, a trend which is reinforced by the inflow of considerable amounts of meltwater from snowdrifts. Later the water feed is reduced and evaporation makes itself felt in the water and isotope balance. Compared with the initial value of the evaporated water (-190 ‰) enrichment by 15 ‰ is found toward the end of the summer. A repeat sample of Feb. 25, 1980 showed a value (-174 ‰) comparable to that of the previous year. The evaporation lakes south of Lake Ozhidaniya and southeast of Lake Zub are in the same range, with -179 ‰ and -171 ‰, respectively. For ^{18}O Lake Krasnoye showed -12.5 ± 0.3 ‰ in 1980.

Subtype No. 3 — Lake Zerkalnoye: This type of lake is intermittently fed from temporary snow and ice fields. In extreme cases the inflow is limited to a few weeks, and then water often evaporates completely during the polar summer. Very small lake basins may remain dry for several years. The degree of evaporation is reflected also in the isotope values. Deuterium enrichment begins at -200 ‰ and reaches a peak at ca. -145 ‰ shortly before the lake disappears. The situation is even more striking in the case of ^{18}O, with -18.5 ‰ (initial value) and -8 ‰ (final value). *Schütze* et al. [13] made a more detailed study of Lake Zerkalnoye.

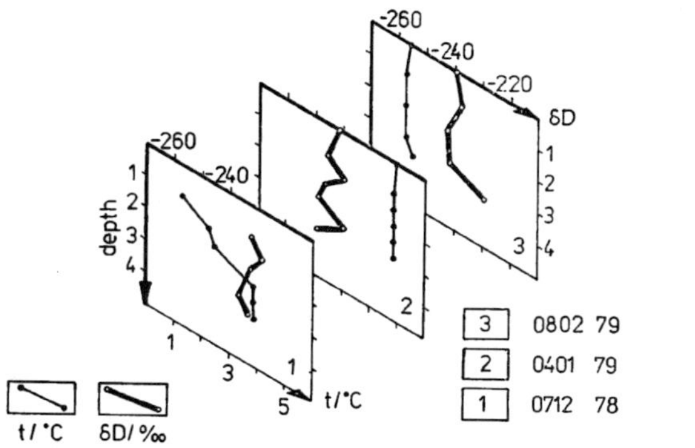

Fig. 4. δD-variations in large and shallow exaration lakes (type Lake Zub)

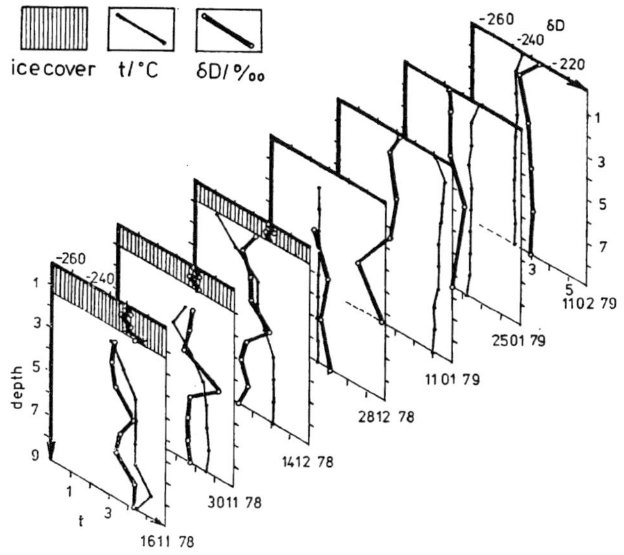

Fig. 5. δD-variations in lakes of medium-size and depth (type Lake Pomornika)

2.2.2.2. D and ^{18}O variations in large and shallow exaration lakes. Large and shallow lakes are frequent in the Schirmacher Oasis. The model selected was Lake Zub which is 0.5 km² in area and has a maximum depth of 5 m. This makes it rank about one or two orders of magnitude above the very small lakes. A lake like this does not freeze down to its floor in winter. On the other hand, the ice cover disappears completely even in an average summer. Fig. 4 shows the deuterium variations and the results of temperature measurements. No clear stratification is found for any measuring date. The δD values are in the range of the input (-230 to -240 ‰). The first sampling indicates a clearly warmer body of water at depths between 2.5 and 4 m, while measurement No. 2 approximately shows the peak of heating at the height of the polar summer. Complete water circulation ensures uniform cooling until ice forms. The ice cover was sampled on Oct. 17, 1979 (167 cm). The δ^{18}O value was -28.0 ± 0.3 ‰ on an average. There is no clear effect from evaporation since even in cool summers the feed and discharge is so heavy that the residence times are less than one month.

2.2.2.3. Isotope variations in exaration lakes of medium depth. Not all of the larger exaration lakes are as shallow as Lake Zub, and many are between 8 and 12 m deep. From a limnological point of view, therefore, one may expect different conditions, such as stratification. Altogether seven determinations were made at two-week intervals taking Lake Pomornika as a model, and the results are shown in Fig. 5. The deuterium variations indicate temporary stratification under the ice cover. The measurement made on Jan. 11, 1979 points to the existence of a stratum of fairly depleted meltwater of a specific temperature.

In contrast with the initial condition where a characteristic temperature profile was found in all three measuring series conducted under an ice cover and where the hypolimnion is at 4 °C, there is complete water circulation after the ice has disappeared. The temperature rose from below 3 °C to more than 5 °C as a result of summer heating and dropped again to around 2 °C by Feb. 11, 1979 without reaching the maximum density. These lakes are normally exposed to strong winds, and the result may be complete circulation and cooling without stratification and a density maximum at 4 °C in the hypolimnion, even in deeper lakes according to Forel [8] temperatures of 4 °C may occur only accidentally at greater depths in "lower-temperature polar lakes" under the meteorological conditions of the polar region.

The sampling dates (Nov. 16, 1978; Dec. 28, 1978; Jan. 11, 1979 and Feb. 11, 1979) give an approximate seasonal picture of the temperature conditions and isotope variations. Winter conditions prevail at the beginning, but the last measurement may be even more descriptive in this respect, and more investigations are needed. As the lake becomes ice-free full circulation starts with balanced temperatures and a uniform profile for δD. Summer conditions are indicated by maximum water temperatures while circulation is retained, and this is followed by the onset of autumn conditions with steady cooling but without thermal stratification. The thermal behaviour and the variations of δD and $\delta^{18}O$ therefore indicate that this type of lake is similar to the large and very shallow exaration lake. However, the greater depth permits acylic stratification to a wider extent.

Evaporation effects cannot be derived with any certainty from the isotope values. Much of the feed for the lakes is meltwater from the inland glacier (−260 ‰), with a minimum δD value around −300 ‰, which is markedly more negative than the meltwater from the oasis (δD around −220 ‰). The residence time of the water should not be much more than a few weeks during the circulation period (no quantitative feed-discharge studies have been made for the area of the oasis, and the estimated residence time has been derived from the time the lake basins need to fill up until maximum discharge is reached).

2.2.2.4. D variations in deep exaration lakes. There are few exaration lakes in the Schirmacher Oasis with depths of 30 m or more. Detailed studies have been made of Lake Glubokoye which was chosen as the model for deep lakes in the summer of 1978/79 (Fig. 6). It receives water from a number of sources including Lake Pomornika. In the presence of an ice cover the δD curve was unstable, and this indicated meltwater strata of varying D abundance. The initial situation of Nov. 17, 1978 was characterized by a comparatively high proportion of meltwater from

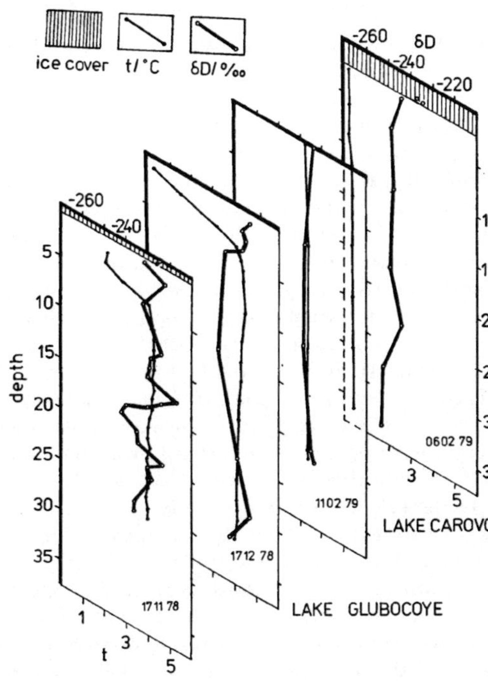

Fig. 6. δD-variations in depth exaration lakes (type Lake Glubokoye) in relation to a medium depth epishelf lake (type Lake Karovoye)

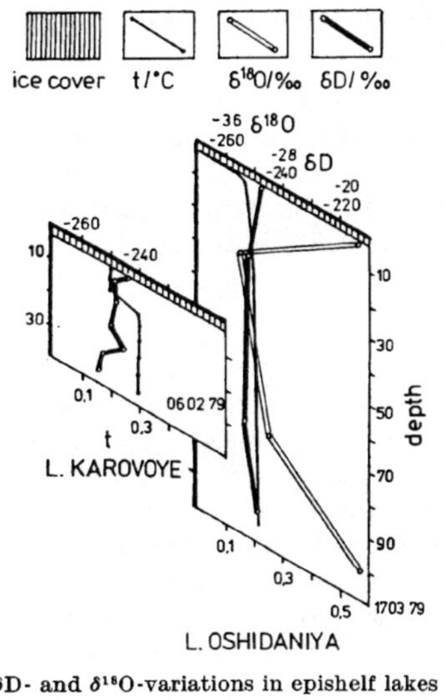

Fig. 7. δD- and $\delta^{18}O$-variations in epishelf lakes (type Lake Ozhidaniya)

snow and ice in the oasis. To the extent that thermal stratification disappeared the conditions were created for full circulation, and the deuterium variations at greater depths were cancelled out. Similar to the preceding type, the body of water cooled down to a temperature below the maximum density by Feb. 11, 1979. Unlike the situation in 1978 the winter conditions in 1979 seemed to give a less marked separation between the epilimnion and the hypolimnion.

By comparing this lake with Lake Karovoye one can illustrate the difference between a lake in the oasis (a rocky area) and an epishelf lake (a rocky glacier area). In contrast with the former, the temperature in the latter is a few-tenths of a degree above zero and no δD variations have been found. In future studies, ^{18}O tests should be used to establish whether stratification is really as week in the deeper bodies of water as is indicated by the δD values.

2.2.2.5. D and ^{18}O variations in the epishelf lakes. The epishelf lakes are the deepest also the strangest freshwater lakes in the Schirmacher Oasis. They were studied from a limnological point of view as early as 10 years ago by the Soviet Antarctic Expeditions. As far as hydrochemistry is concerned they are very similar to the lakes in the oasis. This similarity is of a causal nature since all of these lakes receive meltwater and constitute the last large reservoirs before the meltwater flows into the sea. The temperatures measured by *Simonov* [4] and our own measurements for Lake Karovoye are close to the freezing point (0·2 – 0·3 °C), apart from a slight cooling effect under the ice cover. It can be assumed that this is due to the south edge of the shelf glacier outside the Schirmacher Oasis which is the border toward the sea. The great distances between the sampling points preclude any safe conclusions regarding the isotopic stratification of deep shelf lakes. It is intended to close this gap in future studies. In the case of $\delta^{18}O$ considerable differences seem to exist in Lake Ozhidaniya. The extent to which this can provide information on the origin of individual bodies of water at greater depths will not be known until a detailed analysis has been made of the water balance of the oasis. The δD values are around -240 ‰ and are the most negative compared with the inland lakes of the oasis. For ^{18}O the readings vary between -16 ‰ and -34 ‰. The greatest depletion probably correlates with the water from the glaciers, while enrichment just be attributed to water from snow precipitation or evaporated water from the lakes in the oasis. The following $\delta^{18}O$ values were obtained for two depth samples taken from Lake Zigzag:

9 – 10 m: $^{18}O = -31{\cdot}1 \pm 0{\cdot}3$ ‰; 119 – 120 m: $^{18}O = -30{\cdot}1 \pm 0{\cdot}3$ ‰. In the case of this lake the hydrographical situation excludes higher $\delta^{18}O$ values which result from evaporated water, to a greater extent than in the case of Lake Ozhidaniya which is situated in the central part of the oasis.

2.2.2.6. Lakes on glacial ice. The lakes and ponds on glaciers are among the least studied phenomena of the Schirmacher Oasis because they are farther away from the station and not easily accessible. We were able to take a sample only from one of the large lakes on the inland glacier, Lake Taloye, on Nov. 27, 1979. We drilled through the ice cover which was 289 cm thick. The 33 – 41 cm sample had $\delta^{18}O = -29{\cdot}9$‰ and the 279 – 289 cm sample, $\delta^{18}O = -25{\cdot}1 \pm 0{\cdot}3$‰. The floor was reaches at 4·08 m, and it seems to be jagged. The sample for a depth between 4·28 m and 5·28 m which was from one of the depressions, showed $\delta^{18}O = -25{\cdot}4 \pm 0{\cdot}3$‰.

At a depth of 3·5 m the water temperature was around 0·1 °C, and the temperature for the sample taken from a max. depth of 5·28 m was 0·0 °C. This means that the lakes with the lowest temperatures throughout the year are those completely embedded in the ice.

Following are the results of the δD determination which follow the trend of the $\delta^{18}O$ readings:

Ice sample, 33 – 41 cm: $\delta D = -197$‰
Ice sample, 279 – 289 cm: $\delta D = -193$‰
Water sample, 2·50 – 3·50 m: $\delta D = -199$‰
Water sample, 4·28 – 5·28 m: $\delta D = -210$‰

Whether these results are characteristic of a lake situated on an inland glacier will not be known until comparative values have been obtained for lakes on shelf glaciers.

3. Conclusions

A variety of different types of lakes exist in and around the Schirmacher Oasis, a fact which suggests that the narrow definition of polar lakes should be modified .The differences primarily derive from the wide range of thermal conditions prevailing during the summer months, and these in turn depend on the size and situation of the lake. The smallest ponds (lakelets, "ozerki") showed the greatest thermal variability as was expected, and these properties then virtually "freeze" for eight to ten months when an ice cover forms at the end of summer. His happens with all ice-free bodies of water. A new limnological type is described, the icewall dammed lake. Its thermal behaviour is assumed to be similar to the glacier dammed lakelet. The shallow portions of this transient type seem to be ice-free for the most part during the summer. At least in the case of lakes with a depth of a few meters the formation of an uninterrupted benthonic biological slime has been observed. Particular references is made to the epishelf lakes which have been little studied from a limnological point of view but were described in detail by *Korotkevich* [10] and then by *Kruchinin* and *Simonov* [23]. The authors believe that they belong in the group of inland lakes.

An important element in the hydrochemistry of the lakes situated in the Schirmacher Oasis is the fact that they are a long distance away from the open sea (separated by 80 – 100 km of shelf ice). In contrast with the saline lakes of the genuine coastal oases all bodies of water in the Schirmacher Oasis, including the epishelf lakes connected to the sea, show a low degree of mineralization. The following factors are thought to influence the saline composition of the lakes:

- maritime aerosol input by way of atmospheric circulation,
- the primarily physical weathering of the rocks in the oasis and their leaching by meltwater, and
- the meltwater from the inland galcier with its mineral content deriving from the mountains of Dronning Maud land to be the south.

In addition to hydrochemical and thermal studies tests using deuterium and ^{18}O determination can be employed to establish types of lakes according to the criteria of isotope hydrology. The results of such tests make it possible to explain the role of lakes in the dynamics of the water balance of the Schirmacher Oasis, particularly with regard to the proportions of meltwater flowing in from inland and shelf glaciers and from snow and ice fields in the oasis itself. To a much greater extent than the deuterium variations, studies of the ^{18}O abundance tell us about changes brought about in the water of the lakes by evaporation effects. Work is proceeding to provide ^{18}O analysis on a scale which is equivalent to that of δD studies.

In order to obtain more information from isotope variation studies, it is intended to supplement the feed and discharge investigations designed to give an isotope hydrology balance, with tritium tests covering selected aspects of the water balance of the oasis. From these measurements we expect new approaches to differentiation with regard to the water in the various types of lakes and its balance and dynamics. For future expeditions it is planned to extend the studies to lakes in the mountain cases (Wohlthat Massiv) and on that part of the shelf glacier which receive meltwater from the Schirmacher Oasis. We expect that the results will contribute toward providing a comprehensive description of the Antarctic lakes from the view-point of isotope hydrology. Work is proceeding with the aim of showing the isotope variations in the input to the water balance of the Schirmacher Oasis.

Received October 15, 1982

Literature

[1] *I. K. Rymill, A. Stephenson*, Southern lights. The official account of the British Graham Land Expedition 1934 – 1937, London 1938

[2] *A. Ritscher*, Wissenschaftliche und fliegerische Ergebnisse der Deutschen Antarktischen Expedition 1938/39, 1. Band, Text- und Bilder- und Kartenteil, Leipzig 1942

[3] *A. Ritscher*, Oasen in Antarktika, Polarforschung (1946), (ed. 1948), 70 – 71

[4] *I. M. Simonov*, Oazisy Vostochnoy Antarktidy, Leningrad 1971

[5] ATLAS ANTARKTIKI I, Moskva – Leningrad 1966; II, Moskva – Leningrad 1969

[6] Novolazarevskaya R 31, 32, shelf. ledn. Lazareva R 33, 34, 1:1.000.000, Ministerstvo Morskovo Flota Moscow 1976

[7] Oazis Schirmachera 1:25.000, Ministerstvo Morskovo Flota SSSR, List 1 – 3, Leningrad 1972

[8] *F. A. Forel*, Allgemeine Limnologie, Stuttgart 1901

[9] *M. Hendl, K.-D. Jäger, J. Marcinek*, Allgemeine Klima-, Hydro- und Vegetationsgeographie, Gotha/Leipzig 1978

[10] *E. S. Korotkevich*, Morskye zalivy v oazise Schirmachera na Zemly Korelevy Mod, Inf. Bull. Sov. Antarkt. Eksped. **21** (1960) 10

[11] *E. S. Korotkevich*, Polyarnye pustyni, Leningrad 1972

[12] *G. E. Hutchinson*, A Treatise on Limnology, Vol. I Geography, Physics and Chemistry, New York – London 1957

[13] *H. Schütze, G. Strauch, K. Wetzel*, Stable isotope investigations in Antarctica, Geod. Geoph. Veröff. RI, 1981

[14] *A. Lerman* (Ed.), Lakes. Chemistry, Geology, Physics Springer-Verl. New York – Heidelberg – Berlin (W.) 1978

[15] *I. M. Simonov, K. S. Bonch-Osmolovskaya*, K voprosu o khimizme vod, lda i snega oazisa Schirmachera, Inf. Bull. Sov. Antarkt. Eksped. **72** (1969) 72

[16] *K. L. Blinov*, O. Postuplenii morskikh soley v atmosferu i o znachenii vetra v solevom balanse Kaspiskovo Morya, Tr. Gos. Okeanorg. in-ta, vyp. **15** (27) L. Morskoy Transport 1950

[17] *A. A. Matveyev*, Khimicheski sostav atmosfernikh osadkov po, nablyudeniyam v Mirnom, V. sb.: "Res-ty issled. po progr. MGG, Meteorol. issled." vyp. 5, M. Izd-vo AN SSSR, 1963

[18] Several authors, ZfI-Mitteilungen **37** (1981)

[19] *T. Torii, N. Yamagata, T. Cho*, Antartcic Record (Japan) **28** (1967) 2225

[20] *T. Torii, N. Yamagata, J. Ossaka, S. Murata*, Antarctic Record (Japan) **58** (1977) 116

[21] *O. Matsubaya, H. Sakai* et al., Geochim. Cosmochim. Acta **43** (1979) 7

[22] ... A View on the formation of saline waters in the dry valleys, Memor. Nat. Inst. Polar Res. Spec. Issue no. 13 (1979) 22

[23] *Yu. A. Kruchinin, I. M. Simonov*, Inf. Sov. Antarkt. Eksped. **66** (1967) 12

Isotopenpraxis **19** (1983) 5, S. 153 – 155

Permeation of H_2 and D_2 through Polymers

P. Mercea

(Institute of Isotopic and Molecular Technology Cluj-Napoca, Romania)[1]

The separation possibility of a gaseous H_2-D_2 mixture by selective permeation through polymers is studied. For this, the permeabilities, diffusivities and solubilities of pure H_2 and D_2 at 296 K were measured for six polymers sheets. The dependence of permeabilities on temperature was determined in the range of 260 – 360 K. It was found that the isotopic effect at permeation is strongly dependent on the nature of the polymer and on the temperature. In the end, the possibility of separating a H_2-D_2 mixture by selective permeation through polymers, Pd and microporous glass is discussed.

Keywords

diffusion; hydrogen isotopes; isotope effects; permeability; polymers; temperature dependence

The permeabilities are transfer coefficients in flow of gases through membranes. In the case of the diffusion of a mixture of simple gases through polymer membranes there is normally very little interaction of the permeating components. Thus the separation of such gas mixtures can be predicted when the permeability coefficients for pure components are known. When the separation of a gaseous H_2-D_2 mixture by selective diffusion through polymers is intended it is necesary that, for a given temperature and a pressure gradient, an isotopic effect at permeation exists. Roughly the amount of this isotopic effect will determine the selectivity of the elementary separation effect. In order to establish the optimum separation conditions it is also necesary to determine the dependence of H_2 and D_2 permeabilities and of the permeation isotopic effect on temperature and differential gas pressure at the polymer membrane surfaces.

[1]) Address: R-3400 Cluj-Napoca, P.O.Box 243, Romania

Experimental Method and Results

Six sorts of polymer membranes were investigated: 1) polyethylene terephthalate PET, 2) polyethylene PE, 3) polytetrafluoroethylene TFE, 4) and 5) two sorts of polyvinyl chloride PVC 1 and PVC 2 plastifiated with 25 respectively 35 parts of diocthylphthalate at 100 parts of weight of polymer and finally a cellulose acetate CA membrane. The hydrogen had at least 99% purity while the deuterium, of more than 98% purity was obtained from the reaction between D_2O (99.75% purity) and metallic Na in excess.

To determine the permeabilities P, diffusivities D, and solubilities S, of H_2 and D_2 in the polymer membranes the steady state diffusion method was used. The apparatus and measurements of these coefficients were similar to those already reported [1, 2]. To measure the amount of diffused gas through the membrane the pressure rise in the calibrated downstream volume continously was measured, and graphs of pressure rise against time were ploted for each run. The permeation P, diffusion D and solubility coefficients were evaluated from the linear portion of the graphs and from the time lag Θ, obtained by extrapolating the linear portion to cut the time axis, as follows [1]:

$$P = \frac{dp}{dt} \frac{Vd}{Ap_0} \frac{273}{T} \; (\text{cm}^3\,\text{mm/cm}^2 \cdot \text{s} \cdot \text{atm}) \tag{1}$$

$$D = \frac{d^2}{600\Theta} \; (\text{cm}^2/\text{s}) \tag{2}$$

$$S = \frac{P}{10D} \; (\text{cm}^3(\text{STP})/\text{cm}^3\ \text{polymer})\ \text{at 1 atm} \tag{3}$$

where:

T gas-polymer system temperature, K

p_0 $= p_1 - p_2 \cong p_1$ because $p_2 \ll p_1$, differential gas pressure, cm Hg

Θ time lag, s

$\frac{dp}{dt}$ rate of pressure rise in downstream volume cmHg/s

A membrane area, cm^2

V downstream collecting volume, cm^3

Tab. 1. Diffusion Parameters for H_2 and D_2 in Polymers at 296 K

	Permeability $P \times 10^8$ $\frac{\text{cm}^3\,\text{mm}}{\text{cm}^2\,\text{s atm}}$		Isotopic Effect at Permeation	Diffusivity $D \times 10^7$ $\frac{\text{cm}^2}{\text{s}}$		Solubility $S \times 10^2$ $\frac{\text{cm}^3\,(\text{STP})}{\text{cm}^3\ \text{polymer}}$ at 1atm	
	H_2	D_2	$P_{H_2} \cdot P_{D_2}$	H_2	D_2	H_2	D_2
PET	6·18	5·4	1·15	5·9	4·9	1·04	1·10
PE	78·1	71·1	1·09	58·5	46·6	1·33	1·54
TFE	126·0	106·0	1·19	—	—	—	—
PVC 1	33·8	30·3	1·12	12·3	10·3	2·8	2·95
PVC 2	68·0	64·0	1·06	27·1	24·0	2·5	2·65
CA	63·5	58·3	1·09	38·0	34·0	1·67	1·71

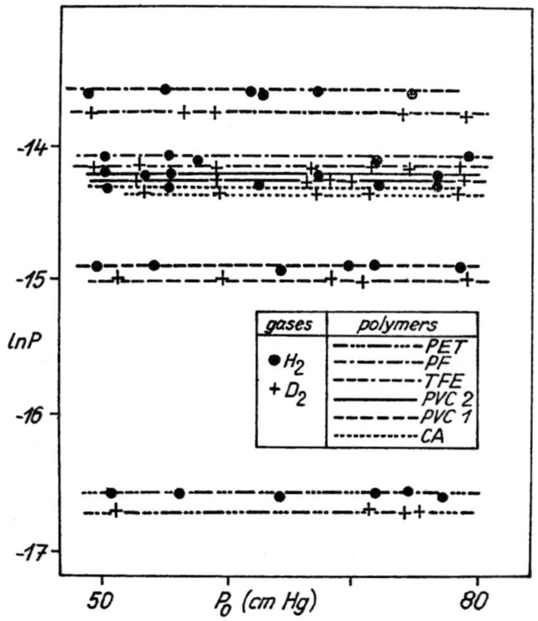

Fig. 1. Differential gas pressure dependence of polymers permeability for H_2 and D_2 at 296 K

For the six polymers the values of P, D and S coefficients and the isotopic effects corresponding to the permeation of H_2 and D_2 are summarized in Tab. 1.

At 296 K for differential gas pressures on the polymer membrane surfaces, ranging from about 50 cmHg to 80 cm Hg the permeabilities for H_2 and D_2 were measured.

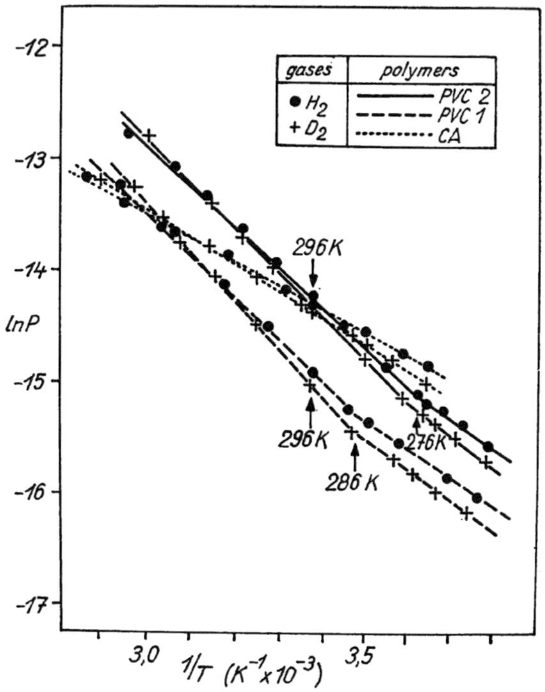

Fig. 2a. Temperature dependence of polymers permeability for H_2 and D_2

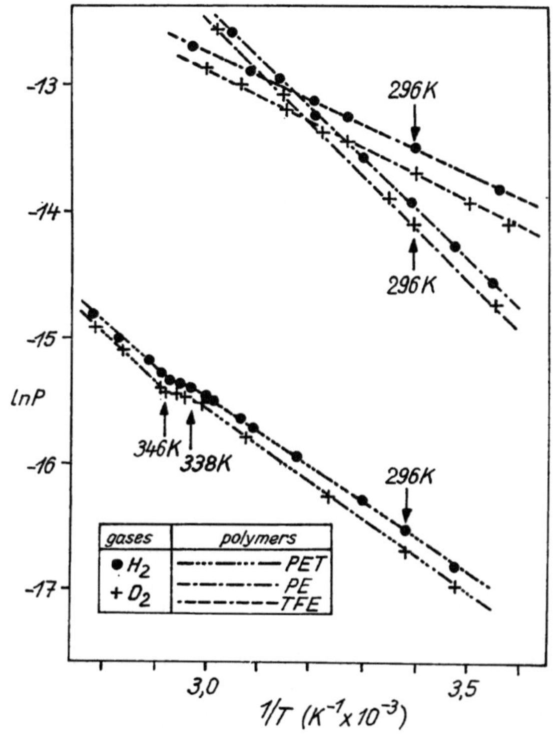

Fig. 2b. Temperature dependence of polymers permeability for H_2 and D_2

The results showed that, within the possibilities of this method, the polymer permeabilities and the isotopic effect at permeation are independent from the differential gas pressure, Fig. 1.

On the other hand, the temperature dependence of the polymer permeabilities for H_2 and D_2 in the range of 260 – 360 K were determined. The obtained results are ploted in Figures 2a and 2b.

Discussion

Tab. 1 shows that at 296 K the permeabilities and diffusivities of H_2 and D_2 through polymers are strongly dependent on the nature of the polymer. According to the activated diffusion model of simple gases through polymers, this fact could be correlated with the existing differences between the number and size of the preexisting holes and the ease of hole formation in those polymers [3]. For the investigated polymers this seems to be true not only for the PET, PE, TFE series but also for the PVC 1 and PVC 2 pair. In the first case differences in chain stiffness, cohesive energy and crystalinity between the three polymers determines differences in permeability and diffusivity. In the second case the higher degree of plastification of PVC 2 causes a larger mobility of the molecular chains, and correspondingly a larger frequency of hole formation which causes a greater permeability and diffusivity for H_2 and D_2.

Tab. 2. Activation Energies of Permeation of H_2 and D through Polymers

	Temperature Range	Activation Energy E_P	
		kcal/mol	
	K	H_2	D_2
PET	285 – 338	5·4	5·7
PET	338 – 346	3·6	3·9
PET	346 – 360	6·6	7·0
PE	283 – 330	7·4	7·6
TFE	280 – 335	4·6	4·8
PVC 1	267 – 286	4·6	5·0
PVC 1	286 – 340	7·9	8·6
PVC 2	260 – 276	5·3	5·5
PVC 2	276 – 325	7·6	8·0
CA	274 – 347	4·3	5·0

The results obtained in this experiments also showed that at 296 K the solubility of H_2 and D_2 is rather small. It is known [4], that in this kind of diffusion experiments the relative error in the solubility measurements is at least five times larger than the error in the permeability measurements. In our experiment this means at least 20% error and therefore the obtained values for H_2 and D_2 are only orientative. However in the case of PVC 1 and PVC 2, where the errors are smaller because of thicker membranes, we were able to conclude that the solubilities of H_2 and D_2 in these polymers decrease with increasing plasticizer amount. This is due to the fact that the addition of plasticizer reduces the number of preexisting holes in which the H_2 and D_2 molecules could be solved by insertion [5].

The temperature dependence of the permeabilities for H_2 and D_2 showed, as expected, that the permeation is a thermally activated process characterized by an activation energy E_P. In the *Arrhenius* plots of Fig. 2a and 2b these energies were calculated from the slope of linear parts of the graphs and the results are summarized in Tab. 2.

As a general rule, in the investigated polymers the activation energies for permeation of D_2 are larger than those for H_2. As a result the isotopic effect at 296 K, reported in Tab. 1, decreases with increasing temperature. There will be a temperature T_a, for each polymer, where the permeability of D_2 equals that of H_2. At this temperature there is no separation between H_2 and D_2 in a diffusion process, that is the mixture of H_2 and D_2 is azeotrope [6]. Above this temperature the permeability of D_2 will be larger than that of H_2. We were able to measure the temperatures T_a for PVC 1, PVC 2 and CA and found them to be about 352, 355 and 350 K respectively. For PET, PE and TFE we extrapolated these temperatures from the ln P versus $1/T$ plots and found them to be about 415·385 and 630 K. These last temperatures were beyond the possibilities of our diffusion apparatus.

Conclusions

At 296 K the selectivity of the studied polymer membranes, for pure H_2 and D_2, vary from about 1·05 to 1·19. This selectivity could be increased by lowering the diffusion temperature. For instance, in the case of CA, where the selectivity increase is the largest, by lowering with 10 K the diffusion temperature the selectivity increases from about 1·09 to 1·14, but simultaneously the amount of diffused H_2 or D_2 decreases by 30%. Comparatively the selectivity of a microporous Vycor glass, at 296 K and upstream pressure $p_0 = 1 - 3$ cmHg is about 1·3 [6] and a Pd membrane at 673 K has a selectivity of about 1·6 [7].

Therefore if a $H_2 - D_2$ separation by gaseous diffusion is intended any of these types of membranes can be used. In order to optimize the membrane and the operating parameters for a $H_2 - D_2$ separation it is required to take into account the interdependency between the selectivity, the amount of diffused gas, the energy efficiency and the costs of the whole process. Although the selectivity of polymers is lower than that of microporous glass or Pd, the separation through polymers could offer some advantages. The membranes are cheaper and when manufactured as hollow fibers they also provide a large diffusion surface and therefore they allow the diffusion of large quantities of gases. At the same time the separation process through polymers does not require special temperature and pressure conditions, being more energy efficient than the diffusion through microporous glass or Pd. These advantages might compensate for the fact that a $H_2 - D_2$ separation through polymers requires more stages.

Received October 8, 1982

References

[1] *F. P. Evans, R. Gibson, C. G. Hutcheson, P. M. S. Jones*, A.W.R.E. Report No. 0 – 45/63, (1963)

[2] *N. N. Li*, Recent Developments in Separation Science, Vol. II Cleveland, Ohio: CRC Press 1972

[3] *J. Crank, G. S. Park*, Diffusion in Polymers, New York: Academic Press 1968

[4] *R. D. Siegel, R. W. Coughlin*, J. Applied Poly. Sci. **14** (1970) 3145

[5] *R. M. Barrer, R. Mallinder, P. S-L. Wong*, Polymer 8 (1967) 321

[6] *E. S. Perry*, Progress in Separation and Purification, New York: Wiley 1968

[7] *S. Tanaka, R. Kiyose*, J. Nucl. Sci. Technol. **16** (1979) 923

Isotopenpraxis **19** (1983) 5, S. 156 – 157

A Hydrocarbon-sensitized Argon Ionization Detector for the Detection of Inorganic Compunds*)

P. Popp, G. Arnold, G. Oppermann

(Zentralinstitut für Isotopen- und Strahlenforschung der AdW der DDR, Leipzig)[1])

Noble gas ionization detectors make it possible to determine inorganic trace components in gases. Among the several modes of operation of these detectors the hydrocarbon-sensitized argon detector is a favourable compromise with regard to the relation of the expense to the attainable detection limit. The suitability of this detector for the determination of H_2, O_2, N_2, CH_4 and CO_2 in pure gases is shown.

Keywords

argon; gas analysis; gas chromatography; penning effect; radiometric gages; sensitivity

1. Introduction

The gas chromatographic determination of inorganic trace components in pure or purest gases is neceassary, among other things, in microelectronics. The most frequently used gas chromatographic detectors (flame ionization detectors (FID) and thermal conductivity detector (TLD)) are not suitable for the solution off this problem because the FID is completely insensitive to inorganic gases and the detection limit of the TLD amounts to 10 – 100 ppm. The determination of permanent gases in the range down to 1 ppm is only possible by means of noble gas detectors. Among the several modes of operation of these detectors the helium detector has the advantage of a very high sensitivity (detection limits lie in a range about 10 ppb) but the disadvantage that the carrier gas (ultra-pure helium) is very costly. Besides, an additional purification is necessary.

With regard to the relation of the expense to the attainable detection limit the hydrocarbon-sensitized argon detector is a favourable compromise.

In this paper the possibilities of such a modified argon detector are described.

2. Theoretical

The argon ionization detector works in the range of nonselfmaintained discharge. The electric field strength is chosen such that the electrons formed by the ionizing radiation are accelerated in such a manner that noble gas atoms can be excited according to

$$Ar + e^- \rightarrow Ar^* + e^-$$

The lowest excitation level of the argon atom is a metastable level and amounts to 11·58 eV.

The metastable excited states formed are able to transmit the excitation energy to other atoms or molecules by collisions of the second kind or by quadrupole radiation. The increase of the ionization current is caused by the *Büttner-Schade* effect

$$Ar^* + Ar^* \rightarrow Ar^+ + Ar + e^-$$

which is effective at high concentrations of metastable excited argon atoms (this means high field strength (about 9000 V/cm) within the ionization region) [1, 2].

The addition of substances with ionization potentials <11·6 eV in low concentrations already at field strengths about 1000 V/cm effects an important increase of the current due to the Penning effect, that means, excited argon atoms, for example, ionize C_8H_{20} molecules according to

$$Ar^* + C_8H_{20} \rightarrow Ar + C_8H_{20}^+ + e^-$$

and field-accelerated electrons may ionize directly:

$$C_8H_{20} + e^- \rightarrow C_8H_{20}^+ + 2\,e^-$$

The addition of gas molecules with ionization potentials >11·6 eV (e.g. H_2, O_2, N_2, CH_4, CO, CO_2) decreases the ionization current because the number of inelastic electron collisions increases and the mean electron energy decreases. The hydrocarbon-sensitized argon detector workes according to this principle.

3. Apparatus

The arrangement for the detection of inorganic compounds is shown in Fig. 1. R 45-argon serves as carrier gas. The injection of the gas samples is accomplished with the three-way valve, the gas chromatographic column is filled with Molecular Sieve 5A or Porapak QS. The molecular sieve column permits the separation of H_2, O_2, N_2, CH_4 and CO, the Porapak QS serves for the separation of the CO_2 of the

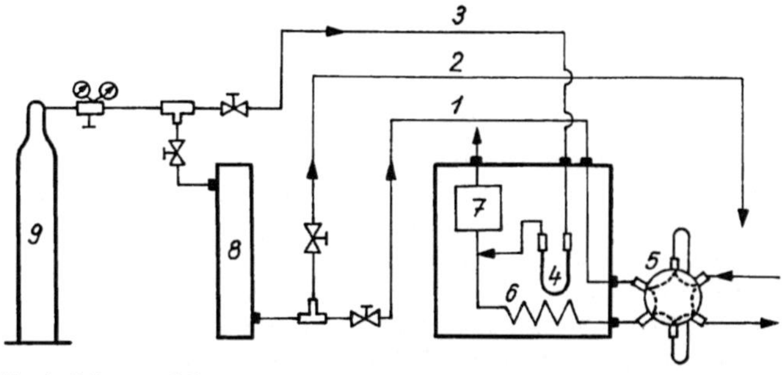

Fig. 1. Scheme of the experimental arrangement

1 – carrier gas; 2 – calibration gas; 3 – addition gas; 4 – diffusion vessel; 5 – gas sampling valve; 6 – g-c-column; 7 – detector; 8 – molecular sieve; 9 – argon R45

*) Presented at the 2nd Working Meeting "Radioisotope Application and Radiation Processing in Industry", Sept. 28 – Oct. 1, 1982, Leipzig.

[1] Address: DDR-7050 Leipzig, Permoserstr. 15

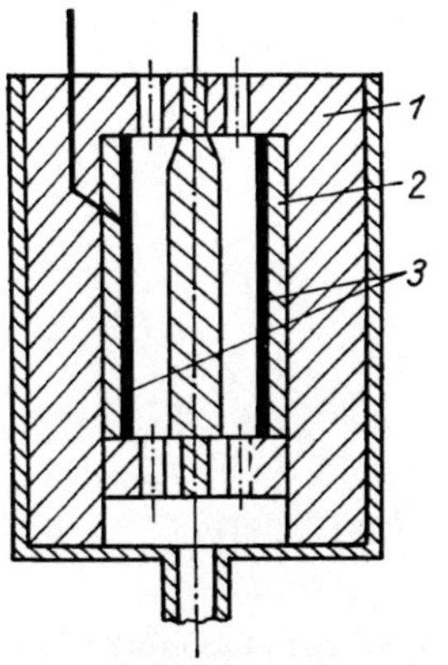

Fig. 2. Scheme of the hydrocarbon-sensitized argon detector
1 – teflon; 2 – stainless steel; 3 – ^{3}H source

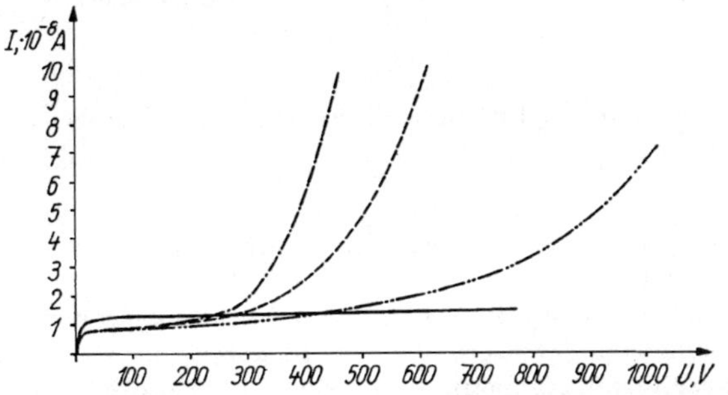

Fig. 3. Dependence of the current-voltage characteristics of the argon detector on the addition of different hydrocarbons
——— argon without addition; –··–··–·· argon with 5 ppm dodecane; – – – – argon with 23 ppm decane; –·–·–·– argon with 100 ppm octane

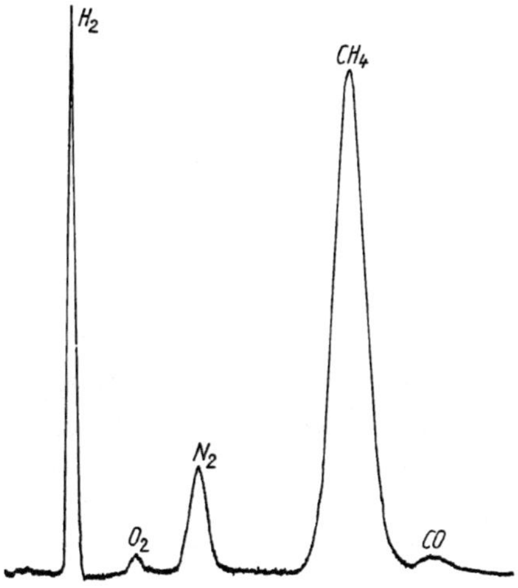

Fig. 4. Chromatogram of a town gas sample
Temperature: 80 °C; Column: stainless steel, 2 m × 4 mm I.D., molecular sieve 5A; Addition compound: 17 ppm decane

air and for the separation of low-boiling hydrocarbons. A parallel operation of both columns is also possible.

An additional carrier gas stream flows over the diffusion vessel filled with C_8H_{18}, C_8H_{20}, $C_{10}H_{22}$, $C_{11}H_{24}$ or $C_{12}H_{26}$ into the detector. The concentrations of C_8-C_{12} hydrocarbons in the ionization room of the argon detector lie between 5 ppm and 100 ppm. An exponential dilution flask permits calibration in the ppm-range. The hydrocarbon-sensitized argon detector shown schematically in Fig. 2 was designed at the Central Institute of Isotope and Radiation Research. The distance between the concentric electrodes is 3 mm, a ^{3}H foil serves as radiation source. The electrodes consist of stainless steel, the insulators of teflon. A high voltage unit (HSVE 20, ZfK Rossendorf) serves as power supply, for the measurement of the current changes a vibrating reed electrometer (Schwing-Kondensator Voltmeter VA-J-52, VEB RFT Meßelektronik Dresden) and a recorder (Labor-Bandschreiber "endim 621.01", VEB Meßapparatewerk Schlotheim) are used.

4. Results and Discussion

Fig. 3 shows the current-voltage characteristics of the argon detector with pure argon and with the addition of 100 ppm octane, 23 ppm decane and 5 ppm dodecane. In the first case the ionization current does not increase considerably as the electric field increases from 200 V/cm (the field, which is necessary to reach the saturation current) up to 3000 V/cm. Due to the mechanisms explained the hydrocarbon admixture effects the current increase at lower field strengths. The analyses were carried out with ionization currents of about $5 \cdot 10^{-8}$ A.

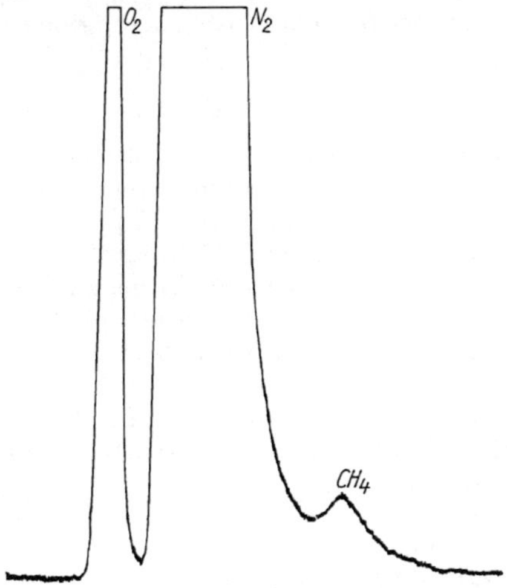

Fig. 5. Chromatogram of an air sample with 50 ppm methane
Temperature: 80 °C; Column: stainless steel, 2 m × 4 mm I.D., molecular sieve 5A; Addition compound: 5 ppm dodecane

Tab. 1. Detection limits of the hydrocarbon-sensitized argon detector (5 ml sample loop)

substance	detection limit, ppm
H_2	2·3
O_2	0·5
N_2	10·5
CH_4	0·5
CO_2	0·2

In Fig. 4 a chromatogram of a town gas sample separated by a Molecular Sieve 5A column is seen and Fig. 5 shows the chromatogram of an air sample containing 50 ppm methane.

The detection limits of H_2, O_2, N_2, CH_4 and CO_2 related to a sample size of 5 ml air are given in Tab. 1.

The values show that using the hydrocarbon-sensitized argon detector it is possible to detect concentrations of inorganic compounds in the ppm-range.

Received September 29, 1982

References

[1] *J. Leonhardt*, Dissertation A, KMU Leipzig 1966
[2] *J. Leonhardt, P. Popp*, Isotopenpraxis **12** (1976) 34

Isotopenpraxis **19** (1983) 5, S. 158 – 160

Measurement of Fluorocarbons with the ECD*)

S. Mothes, P. Popp, G. Oppermann

(Zentralinstitut für Isotopen- und Strahlenforschung der AdW der DDR, Leipzig)[1])

W.-D. Herberg

(VEB Zentrum für Forschung und Technologie Mikroelektronik im VEB Kombinat Mikroelektronik)[2])

Electron-capture detectors enable the high sensitive determination of fluorocarbons and fluorochlorocarbons. It is shown, that the investigation of gas chromatographic retention indices and electron capture coefficients permits the identification of such compounds. Improved ECD modes (coulometric mode, electron attachment spectroscopy) are discussed.

Keywords

electron-capture detectors; etching; gas analysis; gas chromatography; organic halogen compounds; pollution control equipment; radiometric gages; sensitivity

1. Introduction

In the last years the importance of the identification and quantitative detection of fluorocarbons, especially in connection with problems of plasma etching has increased.

Besides, the analysis of freons which possibly effect the degradation of ozone in the atmosphere requires a highly sensitive determination of fluorochlorocarbons.

Both problems can be solved with electron-capture gas chromatography.

In what follows it will be shown that the high selectivity and sensitivity of the ECD in connection with retentiondata can be used in such a manner that unknown fluorocarbons or fluorochlorocarbons are identified and determined in the ppb-range. Furthermore the possibilities of improved ECD-modes (coulometric ECD, electron attachment spectroscopy) for the solution of this analytical problem will be discussed.

2. The use of retention indices and electron attachment coefficients for the identification of unknown components

Relationships between the retention index and other characteristic data of fluorochlorocompounds were examined by *Foris* and *Lehman* [1]. They found that the retention indices decrease with the number of fluorine or hydrogen atoms, but increase with the number of chlorine atoms. The boiling point and molecular weight are proportional to the retention index. Such a correlation was confirmed by our own measurements for fluorochlorocompounds (see Fig. 1). Linear dependences were also obtained for homologous series of fluorocarbons. The knowledge of these correlations enables us to draw conclusions on the boiling points of unknown substances.

The identification of fluoro- and fluorochlorocarbons arising during plasma etching or contributing to air pollution with retention data alone is almost impossible. Only the use of selective detectors guarantees the identification of such compounds.

Electron-capture detectors are especially convenient. In using such detectors it should be noted that the sensitivity for halogen compounds may differ by up to seven orders of magnitude depending on the resonance maximum for dissociative electron capture.

In Tab. 1 relative responses for fluoro- and fluorochlorocompounds given by *Clemons* and *Altshuller* [2] are listed.

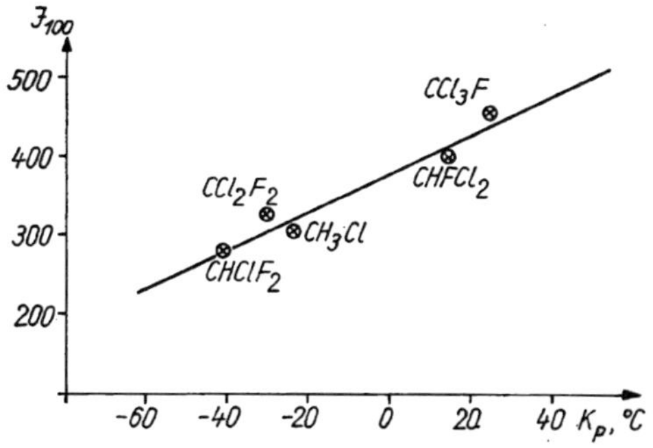

Fig. 1. Variation of retention indices with the boiling point for some halocarbons

Tab. 1. Relative responses of several compounds

Compound	relative response
CHF_3	$<10^{-4}$
CF_4	3×10^{-4}
CH_3CHF_2	9×10^{-3}
C_3F_8	3×10^{-4}
$CF_3CF{=}CFCF_3$	40
$(CF_3)_2C{=}CF_2$	90
$CF_2-CF_2-CF_2-CF_2$ (ring)	30, 40
CHF_2Cl	3×10^{-3}
CF_3Cl	1×10^{-3}
CF_2Cl_3 (Freon 11)	370
CF_2Cl_2 (Freon 12)	9
$C_2F_3Cl_3$ (Freon 113)	50

*) Presented at the 2nd Working Meeting "Radioisotope Application and Radiation Processing in Industry", Sept. 28 – Oct. 1, 1982, Leipzig.

Addresses: [1]) DDR-7050 Leipzig, Permoserstr. 15; [2]) DDR-8080 Dresden, Karl-Marx-Straße

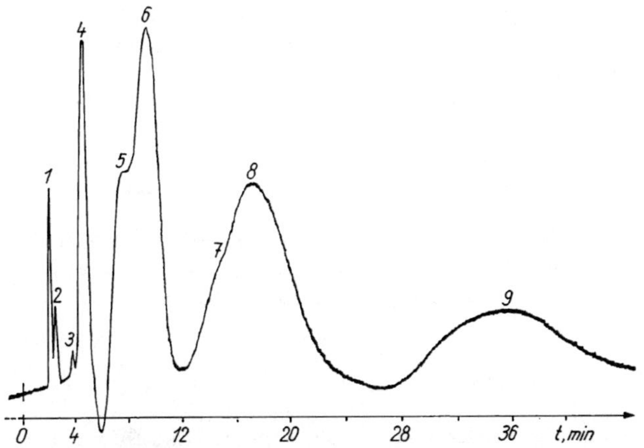

Fig. 2. Chromatogram of waste products of plasma etching processes
Detector: ECD; Column: stainless steel, 1,2 m × 4 mm I.D. filled with Porapak QS; Temperature: 60 °C; Sample: 200 µl waste gas of the CHF_3–O_2-etching
1 — air; 2 — CO_2; 4 — C_4F_6; 5 — c/t-C_4F_8; 6 — iso-C_4F_8

It is shown that the initial products of CF_4 and CHF_3 etching give low responses while the expected waste products of plasma etching, for instance the highly toxic perfluoroisobutene, can be detected in the trace range. Fig. 2 presents a chromatogram of waste products of plasma etching. The components were identified on the basis of correlations of retention indices and boiling points as well as the knowledge of electron attachment coefficients.

The table further shows the highly sensitive detection of freons, especially of freon 11. By use of the ECD freon 11 can be determined down to the ppb- and ppt-ranges without enrichment [3], hence this component is suitable as an indicator of the halogen pollution of the atmosphere.

3. Improvement of the detection limit and selectivity of the ECD

3.1. *Coulometric ECD*

The problem of quantitative analysis is connected with the identification of the components.

If there are no calibration substances and if the electron attachment coefficients or electron attachment cross sections are unknown a calibration is nearly impossible.

In this case the coulometric ECD can be used provided that substances with high electron capture cross sections are present. This mode of ECD operation is appropriate if the electron concentration in the ionization chamber is essentially higher than the concentration of the electronegative gas molecules. For sample substances with high electron attachment cross sections each electronegative molecule is converted into a negative ion and contributes to the decrease of the current. Thus, the sensitivity of the detector is equal to Faraday's number, that means 1 mol of the sample gas produces a quantity of electricity of 96 500 As. A calibration of the detector is not necessary.

Substances determined by coulometric detectors together with their electron attachment cross sections (see *Christophorou* [4, 5]) are summarized in Tab. 2.

Tab. 2. Electron attachment cross sections of several compounds

Compound	σ_{therm} cm²
CCl_4	2×10^{-14}
$(CF_3)_2C{=}CF_2$	6×10^{-15}
$CF_2-CF{=}CF-CF_2$ (ring)	6×10^{-15}
$CF_2-CF_2-CF_2-CF_2$ (ring)	1×10^{-15}
$CF_2-CF{=}CF-CF_2-CF_2-CF_2$ (ring)	2×10^{-14}
$CF_3-C{\equiv}C-CF_3$	6×10^{-15}

3.2. *Electron attachment spectroscopy*

Electron attachment spectroscopy is an analytical method based on the variation of electron attachment cross sections due to changes of the electron energy distribution within the ionization region of an ECD. At the Central Institute of Isotope and Radiation Research preliminary investigations for this spectroscopy were carried out.

A radio frequency voltage was superimposed on the voltage of a direct-current electron-capture detector. This mode of operation makes it possible to get substance-specific "electron attachment spectra".

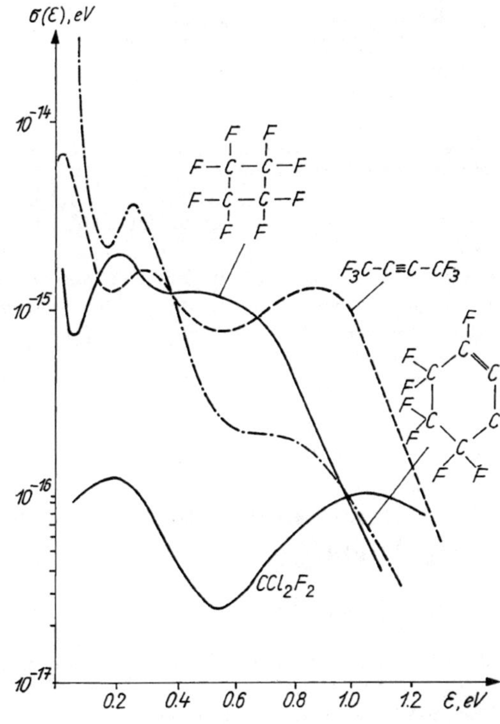

Fig. 3. Dependence of the electron attachment cross section $\sigma(\varepsilon)$ on the electron energy for CCl_2F_2, C_4F_6, c-C_4F_8, C_6F_{10}

Fig. 3 shows such spectra, characterizing the dependence of the electron attachment cross sections on the mean electron energy for the following substances: octafluorocyclobutane, hexafluorobutyne, decafluorocyclohexene and freon 12. It is seen that the $\sigma(\varepsilon)$ — ε-plots are different for the various substances.

If such relations exist for a series of substances it is possible to identify unknown components from such characteristic curve shapes.

Besides, this method enables the sensitivity of the ECD to be essentially enhanced for halogen compounds, if the resonance maxima are higher than the thermal energy.

Fig. 4 illustrates this experimentally established correlation between the electron attachment coefficient K and

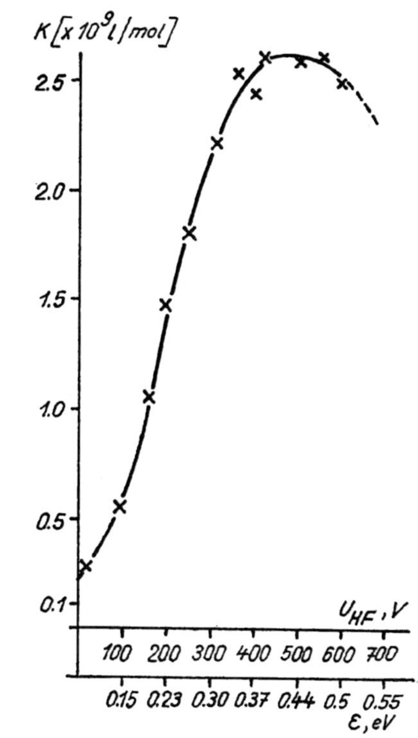

Fig. 4. Dependence of the electron attachment coefficient on the mean electron energy for chloroform

the mean electron energy ε for chloroform [6]. If the amplitude of the radio frequency voltage is chosen such that the mean electron energy is about 0·39 eV, then the coefficient K, which is directly proportional to the sensitivity of the detector, reaches a maximum value.

The energy variation of the electrons in the ECD may be used for the solution of all problems which require highest sensitivity and selectivity.

Highest sensitivity must be guaranteed for the detection of halogen compounds in the atmosphere.

Using this mode of operation it will be possible to determine such compounds quantitatively in air samples without enrichment.

Highest selectivity is necessary for the identification of unknown substances of waste products of plasma etching. The method will be applicable also to this problem.

4. Summary

The highly sensitive detection of fluoro- and fluorochlorocarbons is possible with electron-capture detectors. The dependence of the retention index on the boiling point and the selectivity of the detection caused by differences in the electron attachment cross sections can be used for the identification of unknown components. It has been shown that by means of improved ECD modes (coulometric ECD, electron attachment spectroscopy) the sensitivity and selectivity for such compounds can be increased.

Received September 29, 1982

References

[1] *A. Foris, J. G. Lehman*, Separation Science **4** (1969) 225
[2] *C. A. Clemons, A. P. Altshuller*, Anal. Chem. **38** (1966) 133
[3] *J. Lasa*, Chemia Analityczna **24** (1979) 415
[4] *L. G. Christophorou*, Atomic and Molecular Radiation Physics, Wiley-Interscience, London 1971
[5] *L. G. Christophorou*, Proceedings of the XIIIth International Conference on Phenomena in Ionized Gases 1977, p. 51
[6] *P. Popp, J. Leonhardt, G. Oppermann*, J. Chromatogr. **147** (1978) 127

Isotopenpraxis **19** (1983) 5, S. 160 – 162

Characterization of Coulometric and Hypercoulometric ECD Mode of Operation*)

P. Popp, J. Leonhardt

(Zentralinstitut für Isotopen- und Strahlenforschung der AdW der DDR, Leipzig)[1])

The coulometric mode of operation is given, if each sample molecule passing the ECD is converted into a negative ion and if each negative ion causes a decrease of the ionization current. The sensitivity of the detector in this case amounts to 96 500 C/mol. A hypercoulometric response is given, if fragments from dissociative processes undergo electron capture at several times and if the enhancing effect of space charges is used. In this case the sensitivity of the ECD exceeds the upper limit given by Faraday's number. The operation conditions of both ECD-modes are described.

Keywords

electron capture; electron-capture detectors; environment; gas chromatography; space charge; radiation sources

1. Coulometric mode

Lovelock and coworkers [1] in 1971 first pointed out the usability of the ECD as gas phase coulometer. Such a coulometric mode of ECD operation is given, if the electron concentration *ne* in the detector is essentially higher than the concentration C of the electronegative sample molecules BC, if the electron attachment cross-section is high enough for each molecule BC to be converted into a negative ion (BC^-, B^-, C^-), if each negative ion contributes to the decrease of the current and if the neutral particles (BC, B·, C·) formed by recombination do not take part a second time in the capture process.

*) Presented at the 2nd Working Meeting "Radioisotope Application and Radiation Processing in Industry", Sept. 28 – Oct. 1, 1982, Leipzig.

[1]) Address: DDR-7050 Leipzig, Permoserstr. 15

This is the case if dissociative processes take place because the fragments formed have lower electron attachment cross-sections than the originally existing molecules.

The coulometric mode is accomplished with the pulseECD. Theoretical and experimental studies were carried out by *Lasa* and coworkers [2–4], *Grimsrud* et al. [5, 6] and *Connor* [7]. In order to explain the coulometric mode one regards the change of the concentration of electronegative sample components in time under steady-state conditions:

$$\frac{dc}{dt} = 0 = \frac{Z}{V} - q - \frac{v}{V} c$$

Z, molecules · s^{-1} = number of the sample molecules reaching the ECD per unit time
q, $cm^{-3} \cdot s^{-1}$ = formation rate of negative ions
c, cm^{-3} = concentration of sample molecules
v, $cm^3 \cdot s^{-1}$ = flow rate of the carrier gas
V, cm^3 = volume of the detector

Coulometry means: $q \gg \frac{v}{V} \cdot c$ and

consequently $Z/V \approx q$, the number of the molecules reaching the detector is nearly equal to the number of the negative ions created. If this relation is fulfilled and an optimum detection (all the negative ions recombine) is given then the signal is independent of the sample substance. This means also that the difficult calibration in the range of ppm-ppt-concentrations is not necessary. One obtains

$$\Delta i \cdot t \sim Z \cdot t$$

which means, the charge Q of a peak is equivalent to the total number of the sample molecules. The proportionality

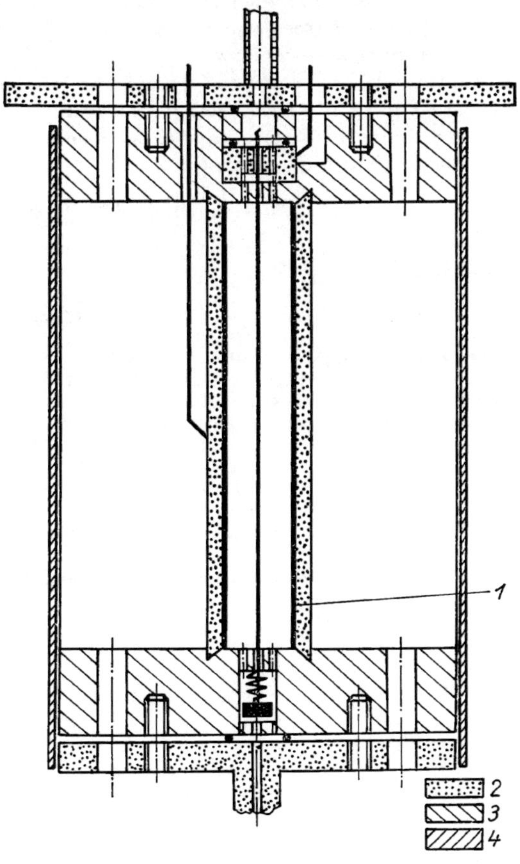

Fig. 1. Coulometric detector with ^3H source
1 – ^3H source; 2 – stainless steel; 3 – teflon; 4 – brass

Tab. 1. Operating conditions and performance of the coulometric detector in the d.c. mode for CCl_4 as test substance

carrier gas	purest nitrogen
gas flow rate	50 cm^3/min
saturation current	3.5×10^{-8} A
basic ionization current under operating conditions	2.5×10^{-8} A
operating voltage	16.2 V
detection limit	1.9×10^{-16} mol/s
sensitivity	1.08×10^5 As/mol
dynamic range	400

constant is *Faraday's* number F so that the relation for coulometry reads

$$Q = F \cdot Z'$$

F *Faraday's* number (96500 As/mol)
Z' = number of moles of the sample passing the detector

Fig. 1 schematically shows the coulometric ECD designed at the Central Institute of Isotope and Radiation Research. The electrodes are arranged concentrically, the inner electrode (the anode) is a wire with a diameter of 0.5 mm, the outer electrode (the cathode) has a 10 mm bore, in which the radiation sources (three ^3H-titanium foils) are placed.

The saturation current with nitrogen as carrier gas amounts to 3.5×10^{-8} A. The increase of the electric field near the wire causes the negative ions formed to be removed rapidly (no occurrence of space charges) and this implies that a coulometric mode is also imaginable under the conditions of the d.c.ECD. The results of first studies with CCl_4 as test substance are presented in Tab. 1. It is seen, that the sensitivity of the ECD for CCl_4 a little exceeds *Faraday's* number.

The calibration was carried out with permeation tubes. The condition $ne \gg C$ is fulfilled, because the number of molecules, reaching the detector per unit time is $10^7 \ldots 10^9$ molecules/s, which is about 2 – 4 orders of magnitude below the number of electrons which reach the anode per unit time (this number amounts to 1.5×10^{11} electrons/s).

2. Hypercoulometric response

Comparing the modes of operation of a commercial ECD *Aue* and *Kapila* [8] pointed out that under pulse conditions the sensitivity of 96500 As/mol, which is to be expected

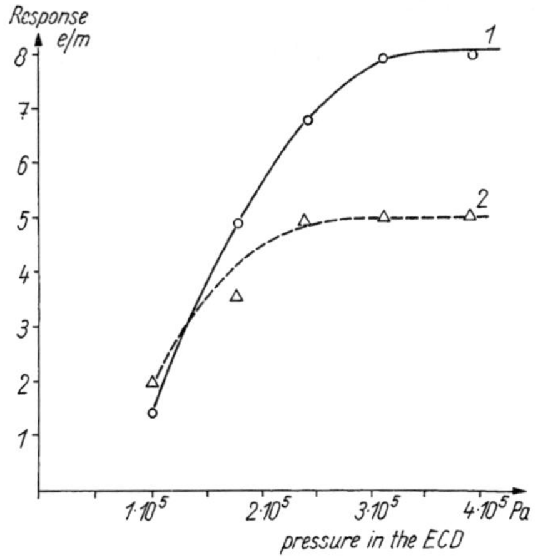

Fig. 2. Occurence of a hypercoulometric effect in a ^3H-ECD in dependence on the pressure for d.c. mode and pulse mode
1 – d.c. mode; 2 – pulse mode

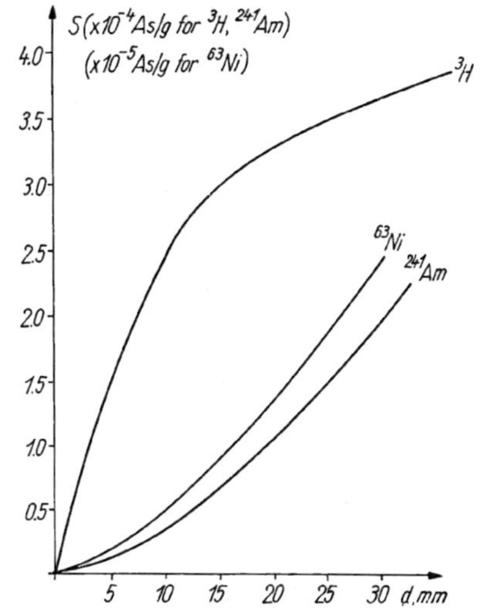

Fig. 3. Dependence of the sensitivity on the electrode distance with negatively polarized ^{3}H, ^{63}Ni and ^{241}Am sources

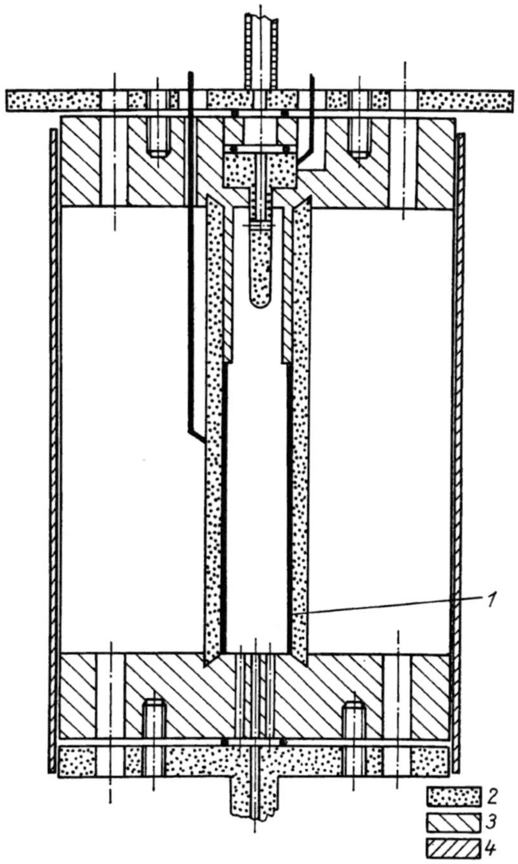

Fig. 4. Hypercoulometric detector with ^{3}H source
1 — ^{3}H source; 2 — stainless steel; 3 — teflon; 4 — brass

for the coulometric response, can be reached and even surpassed but that under d.c. conditions a higher increase (up to a factor 8 as compared with coulometric mode) in sensitivity is possible.

The results of these measurements are given in Fig. 2. The response values represent the quotient of *Faraday's* number and the number of the injected moles (in Fig. 2 marked by e/m).

The authors concluded from this experiment and from the important differences occuring with positive and negative polarization of the radiation source, that space charge effects have a great influence on the processes within the ionization region.

Popp, Leonhardt and *Oppermann* [9], [10] explained the influence of space charges on field and potential in a d.c. ECD with a simple model. They pointed out that the disadvantage of the d.c. ECD, the separation of the charged particles by an electric field, favours the hypercoulometric response. The non-recombining negative ions near the anode form a negative space charge within the range of the unipolar current, which decreases the field and the potential in that region in which charged particles of both kinds of sign exist and the rate of electron-ion recombination increases.

By means of a d.c. ECD with parallel plate electrodes the space-charge caused increase of sensitivity was determined. Fig. 3 shows the increase of sensitivity with increasing electrode distance for three radiation sources (^{3}H, ^{63}Ni, ^{241}Am) and for O_2 as test substance. This increase is space-charge caused because the β-radiation is absorbed nearly totally at a distance of 2 mm (^{3}H) or 10 mm (^{63}Ni, ^{241}Am). It was not possible to obtain a hypercoulometric response by means of the detector with variable geometry because the activity of the sources and consequently the concentration of electrons were too low.

The alterations necessary to change the coulometric detector (see Fig. 1) into a detector with hypercoulometric response are shown in Fig. 4. First experiments with this mode of operation are in preparation.

Received September 29, 1982

References

[1] *J. E. Lovelock, R. J. Maggs, E. L. Adlard,* Anal. Chem. **43** (1971) 1962
[2] *J. Lasa,* Chem. Anal. (Warschau) **24** (1979) 415
[3] *E. Bros, F. M. Page,* J. Chromatogr. **126** (1976) 271
[4] *J. Lasa, E. Bros, I. Sliwka, A. Korus, J. Rosiek,* Institute of Nuclear Physics, Krakow, Report-No. 1029/Ch 1978
[5] *E. P. Grimsrud, S. W. Warden,* Anal. Chem. **52** (1980) 1842
[6] *E. P. Grimsrud, S. H. Kim,* Anal. Chem. **51** (1979) 537
[7] *J. Connor,* J. Chromatogr. **200** (1980) 15; **210** (1981) 193
[8] *W. A. Aue, S. Kapila,* J. Chromatogr. 188 (1980) 1
[9] *P. Popp, J. Leonhardt, G. Oppermann,* ZfI-Mitteilungen **43a** (1981) 127
[10] *P. Popp, J. Leonhardt, G. Oppermann,* J. Chromatogr. **210** (1981) 389

Isotopenpraxis **19** (1983) 5, S. 163–165

^{60}Co Gamma Radiation Damage of Polycrystalline Acetyl Choline Halide Electrolytes

M. M. Abou Sekkina[1]), *N. Hamami, S. A. Abou El-Enein*

Solid samples of acetylcholine halide $[CH_3COOCH_2CH_2 \cdot N(CH_3)_3]X$, where $X = Br, I$ as supplied from (BDH) company have been purified and recrystallized from the appropriate solvent. Extensive studies of infrared absorption spectra and analysis of the X-ray diffraction patterns were carefully conducted before and after exposure to high energy gamma absorbed dose (5×10^8 rd). Results obtained indicated that acetylcholine bromide has higher radiation resistance than acetylcholine iodide and this was discussed on the basis of bond strength and electronegativity differences. These could be also discussed in terms of gamma absorbed dose damage and their attendent creation of lattice imperfections in combination with donor defect centres which are liable to occur pronouncedly for the iodide salt. Finally, the IR absorption spectra as well as X-ray diffraction analysis could find their great usefull application as a rapid and sensitive tools for radiation detection at high efficiency using acetylcholine halides polycrystals as detectors.

Keywords

acetylcholine; gamma radiation; radiolysis; X-ray diffraction

Introduction

Chemical reactions wrought by ionizing radiation involve a broad spectrum of liable intermediates; whilst this consideration underlies the complexity of the subject, it also serves to promote intrinsic interest.

Ionization in radiation chemistry refers to the removal of a valency-shell electron from a molecule. The parent molecule-ion can be regarded as a positively charged free radical, and it represents an important intermediate species. The positive molecule-ion is derived most readily from hydrocarbons by electron or charged-particle impact and in many cases it is an intermediate of great reactivity. In this respect, radiation can be considered from the point of view of its effect rather than its characteristics. Several investigators [1, 2] showed that the radiotic oxidation of vinylpolymers can be studied quantitatively by IR spectroscopy by observing the absorption band at 1730 cm^{-1}. *Oloff* et al. [3] studied the radiation damage in solid single crystals of organic materials and their structures are discussed in relation to the radiosensitizing properties. *Abou Sekkina* and *Assar* [4] investigated radiation damage and hydrogen bonding in some aminoanthraquinone dyes. A correlation between the extent of γ-ray damage and the energy of hydrogen bonding was evaluated. But at the present time, there is little quantitative information as to the potential reactivity of excited states for large polyatomic molecules. Unfortunately, detailed studies on the spectral properties of acetylcholine halides in the solid state is nearly absent in the literature.

The major goal of the present study in the solid state considers a full account of the influence of the high energy ionizing radiation on the spectral and structural properties of acetylcholine halide polycrystals, a study which has not been tackled before. Acetylcholine halide $[CH_3COOCH_2CH_2N(CH_3)_3]X$, where $X = Br^-, I^-$, are selected as being electrolytes with different anionic size.

[1]) Author to whom all correspondence to be addressed: Chemistry Department, Faculty of Science, Tanta University, Tanta Egypt

Experimental

Solid acetylcholine halides $[(CH_3COOCH_2CH_2N(CH_3)_3]X$, where $X = Br^-, I^-$ are supplied as pure chemical grade (99·8%). The materials are freshly recrystallized from absolute ethyl alcohol and dried under vacuum over P_2O_5 for 24 h.

The room temperature IR absorption spectra were carried out using the potassium bromide disk technique in the range 4000 – 625 cm^{-1}. A Pye Unicam (Ltd), Cambridge, England SP 1000 spectrophotometer, Cat. No. 614211 was employed.

A Chimatsu, Japan X-ray diffractometer was used. A Cu-K_α radiation and an adjustable slite at the focal point of the monochromator were undertaken. A *Geiger Müller* tube was adjusted at a rate of one degree per minute. The diffraction patterns were recorded automatically at room temperature in the range of $\theta = 0-60°$, before and after exposure to gamma absorbed dose.

The test samples were exposed to an absorbed gamma dose of 5×10^8 rd using γ-radiation from ^{60}Co – gamma cell (activity 250 Ci). The irradiations were performed in air at room temperature (20 °C).

Results and discussion

Fig. 1, 2 show the schematical representation of the room temperature IR absorption spectra of acetylcholine bromide and iodide polycrystals.

Tab. 1 comprises the main characteristic infrared absorption bands and their assignment for acetylcholine bromide and iodide before and after the prementioned absorbed gamma dose.

As a result of the absorbed gamma dose damage, the following could be easily detected in common:

(a) – The absorption bands broaden.

(b) – Some band are nearly completely disappeared.

(c) – A slight band shift, in few cases, may take place towards small wavenumbers. This could be correlated with the formation of decomposition products at the expense of the present (original) material, vibrating at lower energy states.

Tab. 1. The effect of radiation dose on the characteristics of the main IR absorption bands of acetylcholine halide

Material	ν_{C-C} cm^{-1}		ν_{Ester} cm^{-1}		ν_{C-N} cm^{-1}		ν_{C-H} cm^{-1}	
	A	B	A	B	A	B	A	B
Acetylcholine	1465^{s}	1450^{s}	1720^{s}	1700^{s}	1040^{s}	1025^{m}	2950^{m}	—
bromide					1220^{s}	1210^{s}	2970^{s}	—
					1410^{m}	—	1350^{s}	1350^{m}
Acetylcholine	1460^{m}	1450^{s}	1710^{s}	1705^{s}	1030^{s}	1040^{s}	2950	2970^{m}
iodide					1205^{s}	1195^{s}	1350^{s}	1340^{b}
					1400^{w}			1350^{b}

A = Before radiation dose; B = After radiation dose

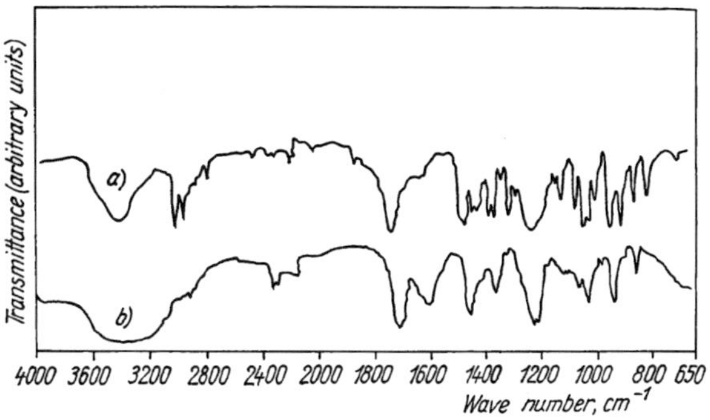

Fig. 1. The room temperature infrared absorption spectra of acetylcholine bromide

a) — before irradiation; b) — after irradiation (5.0 × 10^{6} rd)

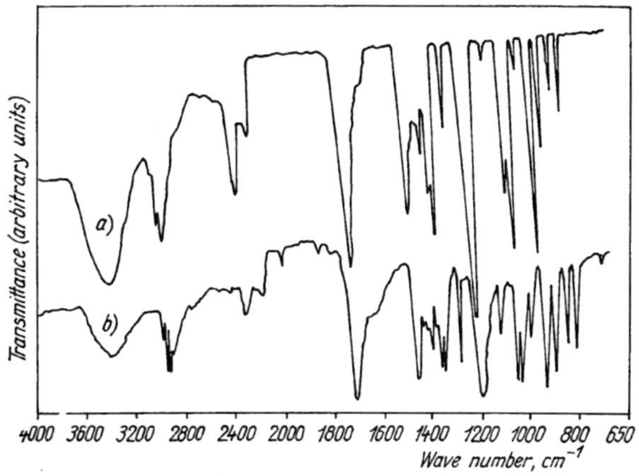

Fig. 2. The room temperature infrared absorption spectra for acetylcholine iodide

a) — before irradiation; b) — after irradiation (5·0 × 10^{8} rd)

Tab. 2. The obtained room temperature X-ray diffraction analysis of acetylcholine bromide

2 °θ	$d_{(°A)}$	I/I_0	2 °θ	$d_{(°A)}$	I/I_0
10·3	8·61140	10	28·6	3·12940	28
15·4	5·76890	100	31·5	2·83760	100
17·4	5·05830	12	32·8	2·73770	6
17·75	5·01010	10	40·1	2·24610	13
18·2	4·88720	91	41·3	2·19150	7
18·9	4·70770	62	42·85	2·11590	93
21·5	4·14410	16	44·80	2·02820	8
21·8	4·08750	42	47·2	1·93090	9
23·5	3·79640	67	48·85	1·86930	41
25·8	3·46230	55	52·4	1·75070	14
26·2	3·41260	19	52·9	1·73540	6
26·9	3·32310	8	53·4	1·72040	9

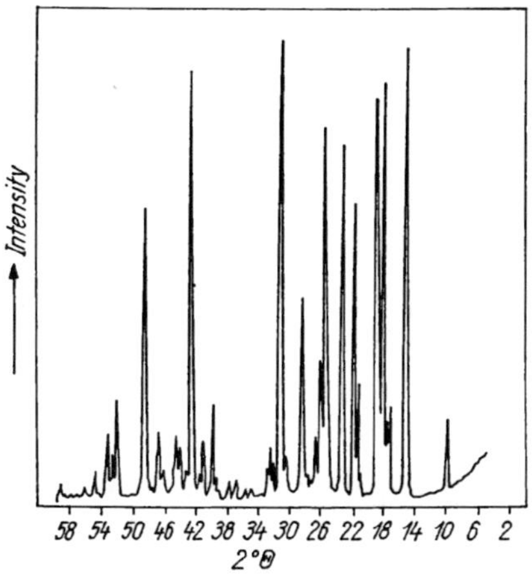

Fig. 3. The obtained room temperature X-ray diffraction pattern of acetylcholine bromide

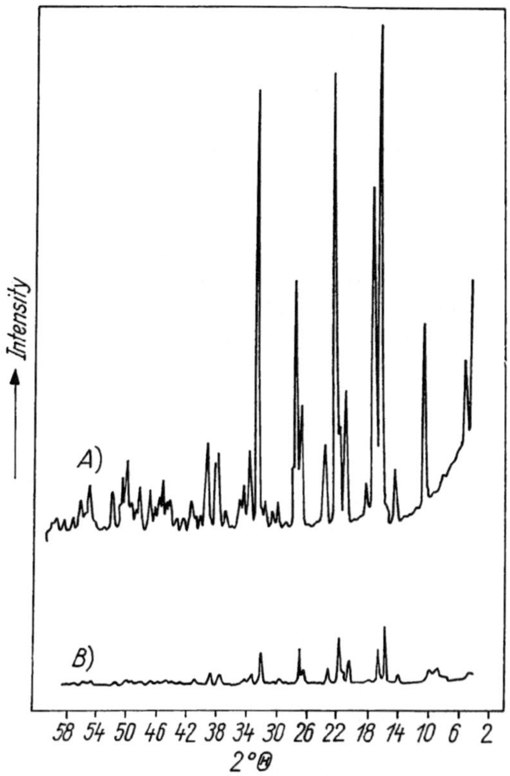

Fig. 4. The obtained room temperature X-ray diffraction patterns of acetylcholine iodide

A) — before irradiation; B) — after irradiation (5·0 × 10^{8} rd)

Tab. 3. The obtained room temperature X-ray diffraction analysis of acetylcholine iodide (A) before and (B) after radiation dose

$d_{(°A)}$		I/I_0		$d_{(°A)}$		I/I_0	
A	B	A	B	A	B	A	B
—	19·19810	—	6	3·33600	3·36150	15	25
—	11·32780	—	8	3·26410	3·27660	31	67
—	10·27560	—	22	2·98670	2·96710	3	7
—	8·66710	—	21	2.76250	2·77830	54	57
8·35630	—	23	—	2·67460	2·68110	10	14
—	6.23340	—	14	2·37390	2·38460	9	18
5·48540	5·57050	100	100	2·29300	2·30810	11	17
5·22970	5·27410	46	60	2·18160	2·18980	3	7
4·86140	4·92510	5	7	2·01130	2·01330	6	8
4·26110	4·30890	17	39	1·93670	1·93990	4	7
—	3·81540	—	22				

Since the main characteristic absorption bands are still retained, thus, the materials individualities are ensured with partial dissociation or recombination which may lead to the formation of large interstitual atom clusters. It can be noted that these changes take place progressively on going from the bromide to iodide salts. Thus, acetylcholine bromide has higher resistance than acetylcholine iodide against the effect of the absorbed gamma dose of 5×10^8 rd. This was most probably correlated with weakening of the ionic bond for the iodide salt due to the increased ionic size and decreased electronegativity of iodine as compared with bromine. This may cause the iodine salt to be less stable (resistant) and hence more liable for radiation damage than for the bromide salt under the same conditions.

For the X-ray diffraction analysis, values of the interplanar spacing d_{hkl} of the *hkl* planes and the relative intensities (I/I_0) for each line were evaluated. Fig. 3, 4 show the room temperature X-ray diffraction patterns of acetylcholine-bromide and -iodide. Tab. 2, 3 comprise their analysis in terms of the interplanar spacings "d" as well as the relative intensities (I/I_0) for each diffraction peak before the absorbed gamma dose (pattern A) and after irradiation (pattern B), whereas acetylcholine iodide was taken as a representative example for the halide salts.

Comparing and matching of the results obtained with those previously given by *Mahayan* et al [5] and *Allen* [6] an orthorhombic crystal form was traced in both cases. Thus, the phenomenon of isomorphism [7] appeared between the bromide and iodide salts of acetylcholine. Crystalline materials ofthis type are isostructural because both crystals are built up of structural units of similar size and shape.

After the absorbed dose, the X-ray diffractograms suffer from the decreased degree of crystallinity indicating great shift to the amorphous phase. This could be explainable on the direct action of the absorbed gamma dose in the formation of decomposition product as well as melting of the material by the energetic gamma rays producing an incorporated glassy phase within the crystalline matrix. The formation of glassy phase reflects itself in the broadening of the IR absorption bands after irradiation. Thus, results of the IR absorption spectra and X-ray diffraction analysis were in conformity.

Conclusion

For further explanation, when gamma rays attack semiconductors of this kind, a large number of ions along with electron dificient regions commonly called holes, are thus produced. The large majority of the freed electrons recombines with the holes within a very short time. Some electrons and also holes may become trapped for long periods especially at lower temperatures. Intrinsic defects (lattice defects or impurities) may act as traping sites. These may reflect themselves in broadening and weakening of the IR and X-ray peaks. As a result, the irradiated materials will have regions with excess trapped electrons called electron trap centres namely at negative ion vacancies, and regions which are deficient in electrons called positive hole centres. These new electronic configuration may cause preferential light absorption (as detected from IR data) at particular wavelengths, and hence are known as "colour centers".

Accordingly, the IR absorption spectra and X-ray diffraction measurements could be used as a sensitive and rapid tools for tracing of gamma radiation damage and the test samples could also be used as gamma ray absorbers in protection or pollution purposes.

Received October 20, 1982

References

[1] *D. S. Ballantine, G. J. Diens*, Monowitz, BNL (1953) 1677
[2] *R. M. Blach, A. Charlesby*, Int. J. Appl. Rad. Isotopes 7 (1959) 134
[3] *H. Olloff, E. Haindi, J. Huttermann, J. Krauss* "Radiat. Research" 80 (1979) 447
[4] *M. M. Abou Sekkina, S. Assar*, Int. J. Appl. Radiat. Isotopes 32 (1981) 847
[5] *A. Mahayan, N. Vijay, S. Sass* and *M. Ronald*, J. Cryst. Mol. Struct. 4 (1974) 15
[6] *K. W. Allen*, Acta Cryst. 15 (1962) 1052
[7] *A. F. Wells*, "Structural Inorganic Chemistry" 3rd ed, Oxford, London 1962

Isotopenpraxis **19** (1983) 5, S. 166–168

Kurze Originalmitteilungen

Charakterisierung von Reinstgraphit durch Neutronenaktivierungsanalyse und Autoradiographie

L. Buchmann, T. Grätsch

(VEB Elektrokohle Lichtenberg)[1])

K. Wagner

(Institut für chemische Technologie der AdW der DDR, Berlin)[2])

Keywords

autoradiography; graphite; impurities; neutron activation analysis; nondestructive analysis; trace amounts

An Graphitmaterialien für die Mikroelektronik werden hohe Anforderungen bezüglich ihrer Reinheit gestellt. Das bezieht sich insbesondere auf solche Elemente, die vom Graphit in das Halbleitermaterial übergehen und die Ladungsträgerkonzentration unkontrolliert verändern können. Ziel der vorliegenden Untersuchung war die quantitative Bestimmung von Spurenverunreinigungen in Reinstgraphit und die Ermittlung von Aussagen über deren Verteilung. Gegenüber dem im VEB EKL angewandten herkömmlichen Verfahren durch optische Emissionsspektralanalyse sollten dabei die Nachweisgrenzen gesenkt und die räumliche Auflösung verbessert werden.

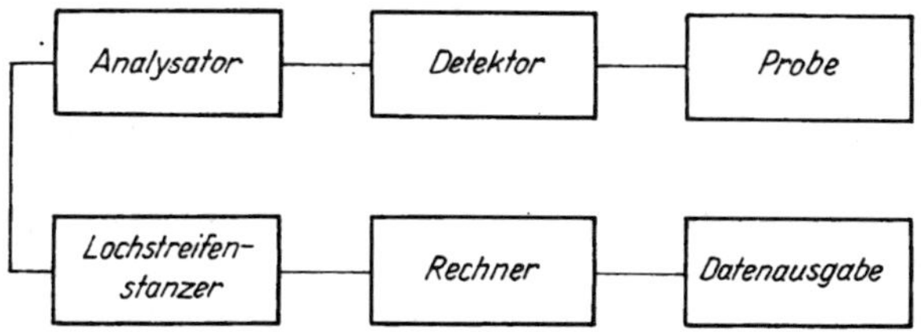

Abb. 1. Schematische Darstellung der Anordnung zur γ-Spektroskopie

Die quantitative Bestimmung erfolgte als zerstörungsfreie Multielement-Neutronenaktivierungsanalyse [1–3] im eichprobengebundenen Direktverfahren, d. h. ohne Trennungen bzw. Anreicherungen (Meßanordnung siehe Abb. 1). Hier wird über die Analyse der Elemente Silber, Arsen, Gold, Kupfer, Natrium, Antimon, Zink berichtet. Es wurden Reinstgraphite mit einem Gesamtaschegehalt unter 10 ppm untersucht. Die Proben (Probenmasse ca. 4 g) wurden 123 h im Kernreaktor mit einem Neutronenfluß von 2×10^{13} cm^{-2} s^{-1} aktiviert. Mit einem Ge(Li)-Detektor und 1024-Kanalanalysator wurden zunächst die kurzlebigen Radionuklide ^{24}Na, ^{64}Cu, ^{76}As, ^{198}Au γ-spektrometrisch gemessen und daraus alle Konzentrationen der betreffenden Elemente bestimmt. Nach 20 Tagen Abklingzeit wurden dann die langlebigen Radionuklide ^{65}Zn, ^{110m}Ag und ^{124}Sb nach derselben Methode bestimmt. Abb. 2 zeigt ein vereinfachtes γ-Spektrum aller untersuchten Radionuklide.

Die Verteilung der Spurenverunreinigungen im Graphit wurde nach zwei Methoden untersucht:

1. Abtragen von Oberflächenschichten und quantitative γ-Spektrometrie des Restkörpers. An zylindrischen Reinstgraphitproben wurden die Gehalte an Silber, Antimon und Zink (^{110m}Ag, ^{124}Sb, ^{65}Zn) vor und nach dem Abdrehen von Oberflächenschichten ebenfalls mit der hochauflösenden γ-Spektrometrie bestimmt und daraus die Konzentration dieser Verunreinigungselemente in der Schicht berechnet.

2. Autoradiographie. Kontaktautoradiographien (auf ORWO AF-4-Film) von Schnittflächen ebenso aktivierter Proben zeigten die Verteilung der Verunreinigungen und wurden photometrisch vermessen (Zeiss-Schnellphotometer G II).

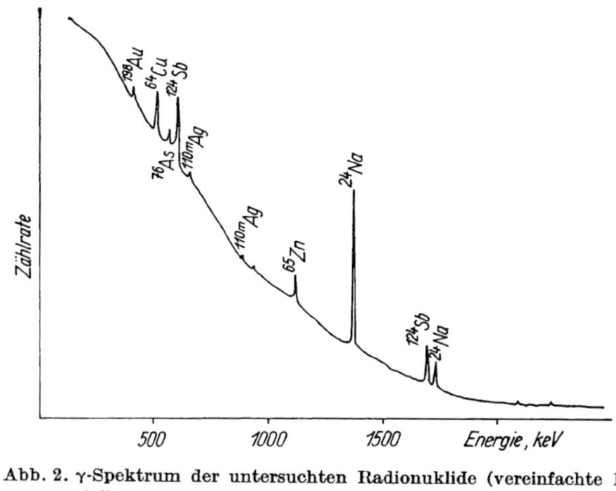

Abb. 2. γ-Spektrum der untersuchten Radionuklide (vereinfachte Darstellung)

Ergebnisse

In je einem Körper der Graphitmarken E 4030 und E 53 wurden nach Entfernung von Oberflächenverunreinigungen die in Tab. 1 zusammengestellten Elementgehalte bestimmt.

Für Konzentrationen, die mindestens eine Größenordnung über der Nachweisgrenze liegen, beträgt der relative Fehler der Radioaktivitätsmessung etwa $\pm 10\%$. Aus der

Anschriften: [1]) 1130 Berlin, Herzbergstraße 128–139; [2]) 1199 Berlin, Rudower Chaussee

Tab. 1. Konzentration einiger Spurenverunreinigungen in ppb (Mittelwerte mehrerer Proben)

	Ag	As	Au	Cu	Na	Sb	Zn
E 4030	0,1	2,0	0,8	2,8	6,3	11,5	12,7
E 53	0,7	2,4	0,5	3,8	6,6	7,8	13,5

Tab. 2. Nachweisgrenzen einiger Elemente in ppb unter den verwendeten Bedingungen

Elemente	Ag	As	Au	Cu	Na	Sb	Zn
Nachweisgrenzen in ppb	10^{-1}	10^{-3}	10^{-4}	10^{-3}	10^{-3}	10^{-2}	10^{-1}

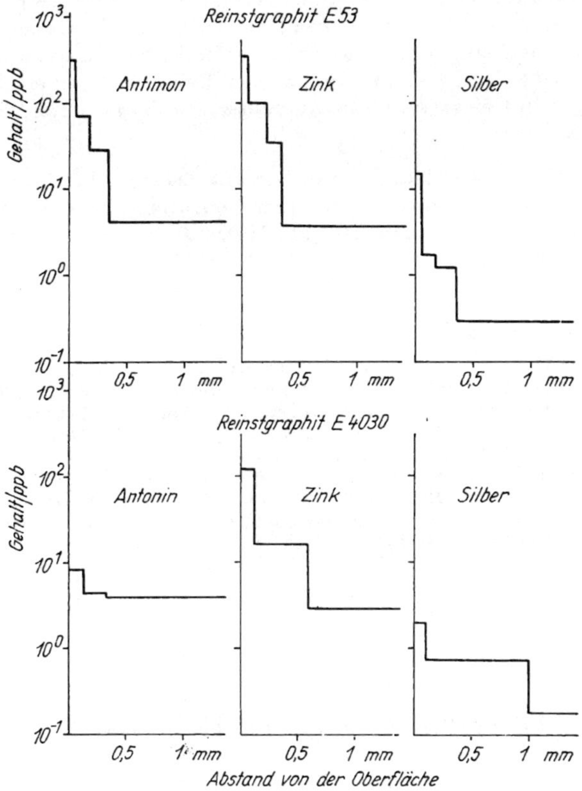

Abb. 3. Durch Schichtabtragung γ-spektrometrisch bestimmte Verteilung der Verunreinigungen Sb, Zn und Ag in den Graphitmarken E 4030 und E 53 (Einzelproben)

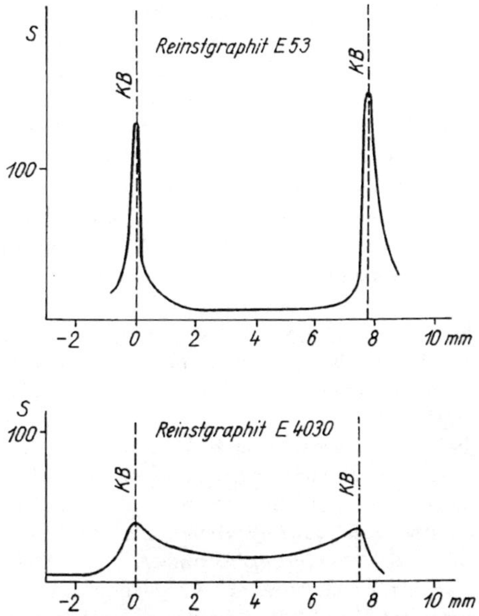

Abb. 5. Photometrische Auswertung der Autoradiographien von Verunreinigungen in Graphitkörpern: Schwärzung als Funktion des Abstandes (Schwärzungsmaxima entsprechen Körperbegrenzung)

gemessenen Impulszahl im entsprechenden Peak des γ-Spektrums lassen sich unter der Annahme, daß 2×10^{-3} Impulse/s noch nachweisbar sind, für die bestehende Anordnung die in Tab. 2 angegebenen Nachweisgrenzen größenordnungsmäßig abschätzen. Diese Nachweisgrenzen stimmen mit Angaben anderer Autoren für Messungen ohne Störungen [4] weitgehend überein, wenn auf die veränderten Aktivierungs- und Meßbedingungen umgerechnet wird. Für die Cu-Bestimmung wurde die Gammalinie bei 511 keV unter Berücksichtigung ihres Abklingens verwendet.

Für diese Reinstgraphituntersuchungen wurden mit der vorher zur Verfügung stehenden optischen Emissionsspektralanalyse nur Nachweisgrenzen von 100 ppb erreicht. Mit der Neutronenaktivierungsanalyse können demgegenüber also einige für die Mikroelektronik relevante Verunreinigungselemente in Reinstgraphit noch bei um fünf Größenordnungen kleineren Konzentrationen bestimmt werden.

Die Verteilung der Spurenverunreinigungen wurde ebenfalls in Reinstgraphitkörpern der Marken E 4030 und E 53 bestimmt. Verfahren 1 (stufenweises Abtragen von Oberflächenschichten, γ-Spektrometrie) erlaubt eine element-

a b c

Abb. 4. Autoradiographien zylindrischer neutronenaktivierter Graphitkörper (Durchmesser ca. 8 mm)
a – Grundfläche; b – Querschnitt; c – Längsschnitt

spezifische Bestimmung und erfolgte für die Elemente Silber, Antimon und Zink (Abb. 3). Die Reinstgraphitproben wurden durch mechanische Bearbeitung auf eine für die Neutronenaktivierung geeignete Form gebracht. Zu den aus Abb. 3 gewonnenen Aussagen sollen hier nur einige kurze Bemerkungen gemacht werden:

Durch Stoffübergang bei der Bearbeitung ergibt sich eine vergleichsweise hohe Konzentration von Verunreinigungen an der Körperoberfläche aller Proben. Die Gehalte an Verunreinigungen werden erst ab etwa 0,3 bis 0,5 mm Abstand von der Oberfläche konstant und sind dort für Zn und Sb beträchtlich niedriger als in Tab. 1 angegeben, weil der Einfluß der hohen Konzentrationen in der Oberflächenschicht dort nur teilweise durch mechanische Dekontamination beseitigt wurde.

Die Ergebnisse werden durch Verfahren 2 (Autoradiographie) bestätigt. Autoradiographien, die auf diese Weise erhalten wurden, sind in Abb. 4 wiedergegeben. Um die als Schwärzungsmaxima sowohl visuell auf den Autoradiographien als auch bei der photometrischen Auswertung (Abb. 5) scharf aufgezeichneten Körperbegrenzungen sind „Schwärzungshöfe" zu finden, die von der hohen Reichweite der β-Strahlung in Luft herrühren, aber die Auswertung der Kontaktautoradiographie des Objektes nicht stören. Bei den verwendeten feinkörnigen und nach der Reinigung mechanisch bearbeiteten Graphitproben befinden sich die Verunreinigungen auch nach dieser Auswertung hauptsächlich in einer dünnen Oberflächenschicht.

Diese beiden Methoden zur Bestimmung des Verunreinigungsprofils bilden keine Alternative, sondern sie ergänzen sich. Bei etwa gleicher räumlicher Auflösung gestattet das Verfahren 1 eine quantitative Bestimmung für einzelne Elemente. Der Zeitaufwand für die Probenbearbeitung ist jedoch relativ hoch. Wegen der für die gesamte Meßreihe nötigen Dauer eignet sich das Verfahren vor allem für langlebige Isotope. Die Autoradiographie (Verfahren 2) bietet bei vergleichsweise geringem Aufwand einen schnellen Überblick, es können vor allem viele Proben gleichzeitig autoradiographiert werden, und erfaßt vor allem β-strahlende Isotope. Die Methode ist nicht elementspezifisch und läßt eine quantitative Bestimmung nur dann zu, wenn wenige β-Strahler die Schwärzung verursachen und sich in ihren Halbwertszeiten und β-Energien genügend unterscheiden [5, 6].

Schlußbemerkungen

Für die Spurenbestimmung von Elementen in Reinstgraphit ist die Aktivierungsanalyse mit thermischen Neutronen wegen ihrer extrem hohen Empfindlichkeit ausgezeichnet geeignet, nicht nur für alle hier genannten, sondern auch für viele andere Elemente. Diese gute Eignung ergibt sich auch deshalb, weil das Matrixelement Kohlenstoff nicht stört und weil Verunreinigungen, die nach der Neutronenaktivierung aufgebracht werden, ebenfalls die Analysen nicht verfälschen. Deshalb konnten auch die Elementverteilungen in Reinstgraphitproben in der hier dargestellten Weise bestimmt werden.

Für die tatkräftige Unterstützung bei der Durchführung der Experimente möchten sich die Autoren beim Strahlenschutzbeauftragten des IcT, Herrn *F. Brutschin*, herzlich bedanken.

Eingegangen am 7. 9. 1982

Literatur

[1] *L. Buchmann, T. Grätsch, K. Wagner*, Charakterisierung von Reinstgraphit durch Neutronenaktivierungsanalyse, Poster: XXXIII Berg- und Hüttenmännischer Tag, Freiberg, 22. 6. bis 25. 6. 1982

[2] *J. J. Rowe, E. Steinnes*, J. Radioanalyt. Chem. **37** (1977) 849

[3] *C. Block, R. Dams*, Analytica Chimica Acta **68** (1973) 11

[4] *M. L. Verheijke*, Philips Techn. Rundschau **34** (1974/75) 334

[5] *K. Wagner, H. Lucke, F. Brutschin*, Isotopenpraxis **15** (1979) 242

[6] *K. Wagner, R. Langenstraßen, S. Ludwig, M. Menzel*, Schmierungstechnik **12** (1981) 74

Isotopenpraxis **19** (1983) 5, S. 168–170

Synthese des 3-Pyrrolidinosulfonylesters von 17α-Ethinyl-[9α, 11α-³H]-estradiol

J. Römer

(Zentralinstitut für Kernforschung Rossendorf)[1]

S. Schwarz, G. Weber

(VEB Jenapharm, Jena)[2]

Keywords

steroids; synthesis; tritium compounds

In einer früheren Arbeit [1] berichteten wir über 9α, 11α-tritiummarkierte Estradiolderivate. Wir beschrieben die Synthesen von [9α, 11α-³H]-Estradiol (1a), 17α-Ethinyl-[9α, 11α-³H]-estradiol (1b), dessen 3-Isoprophylsulfonat (1c) und dessen 3-[N,N-Diethylamino]-sulfonat (1d). In der vorliegenden Arbeit wird die Synthese eines anderen pharmakologisch interessanten 3-Sulfonylesters von (1b) erläutert. Es handelt sich um 17α-Ethinyl-[9α, 11α-³H]-estradiol-3-pyrrolidinosulfonat (1e) mit hoher spezifischer Aktivität.

(1)

	R^1	R^2
a	H	H
b	H	C≡CH
c	$i\text{-}C_3H_7\text{-}SO_2$	C≡CH
d	$Et_2N\text{-}SO_2$	C≡CH
e	$C_4H_8N\text{-}SO_2$	C≡CH

Anschriften: [1]) 8051 Dresden, PF 19; [2]) 6900 Jena

Ergebnisse und Diskussion

Wir hydrierten 9(11)-Dehydroestronacetat (2) nach Vorschrift [1] katalytisch mit T_2. Dabei fiel das [9α, 11α-^{3}H]-Estronacetat (3a) mit einer spezifischen Aktivität von 2046 GBq/mmol und einer radiochemischen Reinheit von 94% an.

Die anschließende Ethinierung zum 17α-Ethinyl-[9α, 11α-^{3}H]-estradiol (1b) führten wir nach *Beumel* und *Harris* [2] mit überschüssigem Ethinierungsreagens durch und verfolgten die Umsetzung mittels Radiodünnschichtchromatographie (Radio-TLC). Günstig sind kurze Reaktionszeiten bis zu 2 Stunden. Wir benötigten 5 Stunden, um eine 80%-ige Umwandlung zum Ethinylestradiol (1b) zu erzielen. Das drückte sich sofort in einer größeren Abnahme der spezifischen Aktivität aus, weil im stark basischen Reaktionsmedium Aktivität aus der semistabilen benzylischen 9α-Position [1] ausgetauscht worden war. Wir registrierten deshalb für (1b) nur noch eine spezifische Aktivität von 1428 GBq/mmol.

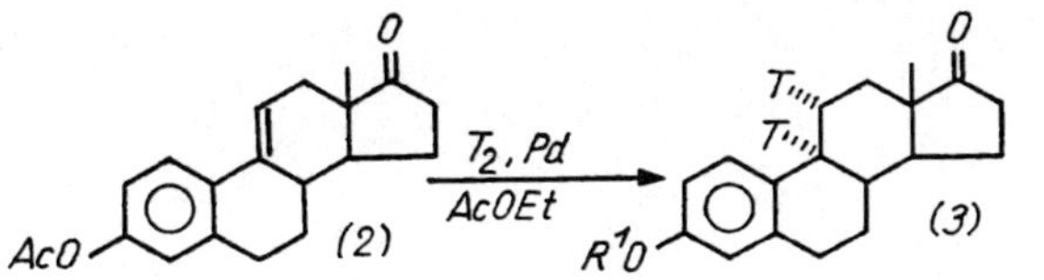

	R^1
a	Ac
b	H
c	$C_4H_8N\text{-}SO_2$

Alle Werte für die Abnahme der spezifischen Aktivität im Ethinierungsschritt, die wir bisher bestimmten, sind in Tab. 1 zusammengestellt. Man erkennt, daß sich der Wert für 5 Stunden Reaktionszeit einwandfrei in die logarithmische Abnahme der spezifischen Aktivität einfügt.

Tab. 1. Abnahme der spezifischen Aktivität von [9 α, 11 α-^{3}H]-Estra-1, 3, 5(10)-trienen in Abhängigkeit von der Aufenthaltsdauer im stark basischen Ethinierungsmedium

Zeit, h	spezifische Aktivität von (1b) in % gegenüber der Ausgangsverbindung (3)	Literatur
2	77,7	[1]
3	80,6	[1]
5	69,8	diese Arbeit
16	52,3	[1]

Die Veresterung von (1b) mit Pyrrolidinosulfonylchlorid zum 3-Pyrrolidinosulfonat (1e) führten wir phasentransferkatalysiert durch [3]. Mittels Radio-TLC wiesen wir nach, daß quantitative Veresterung bereits nach 30 min stattgefunden hatte. Das klebrige Rohprodukt wurde durch präparative TLC an Kieselgel gereinigt.

Im Ethinierungsmedium wurde (3a) vorerst zum [9α, 11α-^{3}H]-Estron (3b) verseift. Weil die anschließende Ethinierung nicht quantitativ verlief, wurde restliches (3b) ebenfalls verestert. Das gebildete [9α, 11α-^{3}H]-Estron-3-pyrrolidinosulfonat (3c) ist wesentlich unpolarer als (1e) und konnte deshalb durch präparative TLC abgetrennt werden. Radiogramme vor und nach der phasentransfer-katalysierten Veresterung sind in Abb. 1 und 2 dargestellt.

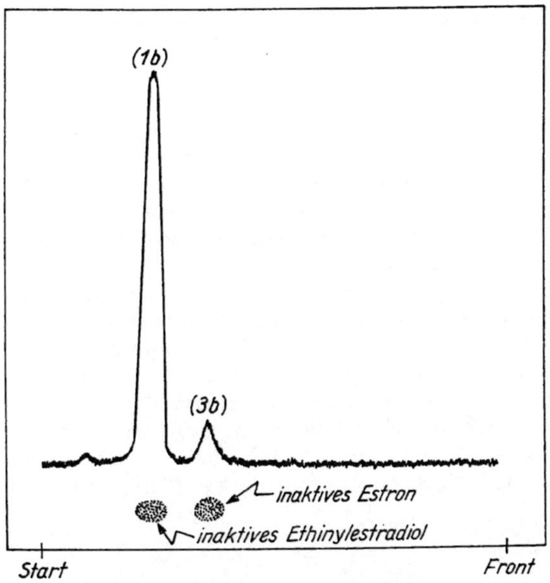

Abb. 1. Radiogramm nach der fünfstündigen Ethinierung des [9α,11 α-^{3}H] Estrons (3b) zum 17α-Ethinyl-[9α, 11α-^{3}H] estradiol (1b)
Laufmittel: Benzen/Chlf./MeOH (80:15:5); R_f (1b) = 0,24; R_f (3b) = 0,36

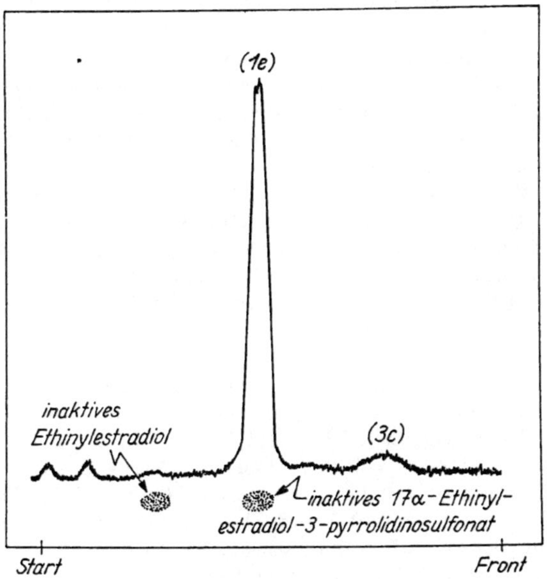

Abb. 2. Radiogramm nach der phasentransfer-katalysierten Veresterung von (1b) zum 17α-Ethinyl-[9α, 11α-^{3}H] estradiol-3-pyrrolidinosulfonat (1e)
Laufmittel: Benzen/Chlf./MeOH (80:15:5); R_f (1e) = 0,48; R_f (3c) = 0,75

Experimentelles

Geräte, Chemikalien, Hilfsmittel. Alle aktiven Versuche wurden in Boxen durchgeführt. Die Hydrierung mit T_2 (Import aus der UdSSR) erfolgte in der Rossendorfer Markierungsapparatur [4]. Zur Bestimmung der spezifischen Aktivitäten wurde das Flüssigkeitsszintillationsspektrometer LS-233 (Beckman, USA), zur Auswertung von Radio-TLCs der Scanner II (Berthold-Friesecke, BRD) verwendet.

Als Laufmittel für präparative Kieselgelplatten (Merck, 2 mm, BRD) oder analytische Silufolplatten (Kavalier, ČSSR) diente Benzen: Chloroform: MeOH = 80:15:5. Silufolplatten wurden nach der Entwicklung zur Sichtbarmachung der Flecke mit Chloramin-T-Schwefelsäure [5] besprüht und erwärmt.

Die Reaktionsstufen (2) → (3a) → (3b) → (1b). Diese Reaktionen wurden bereits in [1] ausführlich beschrieben. In der vorliegenden Synthese starteten wir mit 170 μmol (2) und isolierten 47,6 μmol (1b); das sind 28%.

17α-Ethinyl-[9α,11α-^3H]-estradiol-3-pyrrolidinosulfonat(1e). In einem Rundkolben mit Rührmagnet wurden (1b) (13,5 mg = 45 μmol, A_s = 1428 GBq/mmol) und Benzen (3 ml) als Suspension vorgelegt. Nach Zugabe von Triethylbenzylammoniumchlorid (1 mg), Pyrrolidinosulfonylchlorid (1 Trpf.) und NaOH (konz., 1 Trpf.) wurde in einem Bad (90 °C) 30 min unter Rühren und Rückfluß zum Sieden erhitzt. Nach Zugabe von Benzen (5 ml) wurde mit NaCl-Lösung (gesättigt, 10-ml-Portionen) alkalifrei gewaschen. Gefriertrocknen der benzenischen Phase lieferte ein gelbes, klebriges Produkt (21,8 mg), das auf einer präparativen Kieselgelplatte chromatographiert wurde. Die Zone des 3-Pyrrolidinosulfonats (1e) wurde mit Benzen/Ether (1:1) (15 ml) portionsweise erschöpfend eluiert. Gefriertrocknen des Eluats lieferte 8,0 mg (= 18,4 μmol) (1e) mit 1258 GBq/mmol und einer radiochemischen Reinheit von 99,5%. Diese Menge stellt, bezogen auf eingesetztes (2), eine Ausbeute von 11% dar.

Eingegangen am 22. 10. 1982

Literatur

[1] *J. Römer, S. Schwarz, G. Weber,* J. Lab. Comp. Radiopharm. **18** (1981) 1295

[2] *O. F. Beumel, R. F. Harris,* J. org. Chem. **29** (1964) 1872

[3] *S. Schwarz, G. Weber,* Z. Chem. **15** (1975) 270

[4] *Römer, J.* et al., ZfK-Report 445 (1981)

[5] *K. L. Bajaj, K. L. Ahuja,* J. Chromatogr. **172** (1979) 417

Isotopenpraxis **19** (1983) 5, S. 170

Tagungsberichte

„Seminar on High-Dose Dosimetry in Industrial Radiation Processing"

Die Internationale Atomenergie Organisation (IAEA) führte in der Zeit vom 20. September bis zum 1. Oktober 1982 im Nationallaboratorium RISØ/Dänemark ein Seminar über Hochdosisdosimetrie für industrielle Bestrahlungsprozesse durch.

Veranstaltungen dieser Art finden nicht sehr häufig statt, obwohl sich strahlentechnische Prozesse in der Welt — insbesondere in der chemischen Industrie, für die Bestrahlung von Lebens- und Futtermitteln sowie von medizinischen Artikeln usw. — zunehmend durchsetzen.

Da für solche Prozesse eine exakte Dosimetrie, d. h. die möglichst genaue Erfassung der applizierten Strahlendosis erforderlich ist, war das Interesse an dem Seminar international beachtlich. Vertreter aus 18 Mitgliedsländern der IAEA (Australien, VR Bulgarien, ČSSR, Ägypten, DDR, BRD, SFR Jugoslawien, Malaisia, Mexiko, Niederlande, Philippinen, VR Polen, Schweden, Südafrika, VR Ungarn, England, USA und Venezuela) nahmen an dem Seminar teil, dessen Ziel es war, sowohl die theoretischen Grundlagen als auch die experimentellen Möglichkeiten der Hochdosisdosimetrie darzustellen.

In einem seitens der IAEA (Dr. *Nam*, Vertreter der Dosimetry Section — Division of Life Science) und des Nationallaboratoriums Dänemarks (Dr. *Arne Miller*) ausgezeichnet organisiertem Programm wurden die Seminarteilnehmer mit den zur Zeit gängigsten Methoden der Hochdosisdosimetrie vertraut gemacht.

Der Vorteil der Veranstaltung lag darin, daß die Teilnehmer die Möglichkeit hatten, sechs verschiedene dosimetrische Methoden (Kaliumdichromat-, Film-, Chlorbenzol, PERSPEX-, Lumineszenz- und Cersulfat- sowie Fricke-Dosimetrie) selbst experimentell in den Räumen des Beschleunigerdepartments des Nationallaboratoriums RISØ zu praktizieren und damit Fehlerquellen und Meßgenauigkeit des jeweiligen dosimetrischen Systems einschätzen zu können.

Für die Bestrahlungsversuche standen die Bestrahlungsanlagen des Nationallabors RISØ (Elektronenbeschleuniger und Cobalt-60-Bestrahlungsanlagen) sowie eine industrielle Cobalt-60-Bestrahlungsanlage in Roskilde zur Verfügung.

Im Rahmen des Seminares wurde auch das seit 1977 laufende Programm der IAEA „The High-Dose Standardization and Intercomparison for Industrial Radiation Processing Programme of the IAEA", an dem eine Reihe von Mitgliedsstaaten der Organisation beteiligt sind, vorgestellt und diskutiert.

Darüber hinaus fanden die Probleme der Schaffung nationaler dosimetrischer Standards und der personendosimetrischen Überwachung strahlenexponierter Personen breiten Raum in der Diskussion.

Die auf dem Gebiet der Hochdosisdosimetrie gewonnenen Erkenntnisse und Erfahrungen der verschiedensten Mitgliedsländer der IAEA sollen auf einem internationalen Symposium im Jahre 1984 vorgestellt und diskutiert werden.

Weitere Informationen über das Seminar können im Zentralinstitut für Isotopen- und Strahlenforschung der AdW oder im Staatlichen Amt für Atomsicherheit und Strahlenschutz erhalten werden.

M. Bär

Isotopenpraxis 19 (1983) 5, S. 171–177

Informationen

OECD. Gemäß dem kürzlich veröffentlichten Jahresbericht der Kernenergieagentur (NEA) der OECD erreichte der Anteil der Kernenergie an der gesamten Elektrizitätsproduktion im OECD-Raum im Berichtsjahr 13%, was einer Steigerung um 17% gegenüber 1980 entspricht. In fünf Ländern lag der Atomstromanteil höher als 25%, nämlich in Frankreich (38%), Schweden (37%), Finnland (34%), in der Schweiz (29%) und in Belgien (26%). Wie der Bericht festhält, werden sowohl Frankreich als auch Belgien schon 1985 mehr als 50% ihtes Stromes in Kernkraftwerken erzeugen.

Der Zuwachs der installierten Nuklearkapazität um 17% im Jahr 1981 lab stark über demjenigen der Vorjahre. Er hatte 1980 6% und 1979 nur 2,5% betragen. Ende 1981 waren in den OECD-Ländern Kernkraftwerke mit einer Leistung von 134000 MW in Betrieb. Bis 1985 sollten 220000 MW und bis 1990 rund 320000 MW erreicht werden. Diese Entwicklung würde einer Zunahme um 13% zwischen 1981 und 1985 und von 7,8% von 1985 bis 1990 entsprechen. Durchschnittlich müßten dazu in der erstgenannten Periode jährlich 21 neue Blöcke, im zweiten Zeitabschnitt jährlich 20 neue Einheiten den Betrieb aufnehmen.

(SVA-Bulletin)

BRD. Mit seinem Urteil vom 31. März 1982 hatte der Verwaltungsgerichtshof von Baden-Württemberg die Rechtmäßigkeit der ersten Teilerrichtungsgenehmigung für das Kernkraftwerk Wyhl aus dem Jahre 1975 bestätigt. Nunmehr liegt die 568 Seiten umfassende Begründung für das Urteil vor. Darin stellt der Gerichtshof fest, daß die Pläne für das Kernkraftwerk dem heutigen Stand von Wissenschaft und Technik entsprechen und jeder erdenkliche Schutz gewährleistet sei. Ein Bersten des Reaktordruckbehälters sei praktisch ausgeschlossen, ein zusätzlicher Berstschutz daher nicht erforderlich. Kühlmittelverlust-Störfälle könnten heute zuverlässig beherrscht werden, ein Kernschmelzen könne praktisch nicht eintreten, stellt der Gerichtshof fest. Ausführlich werden der Schutz gegen äußere Einflüsse und die Belastung der Umwelt behandelt. Es sei nicht zu erwarten, daß Menschen durch den Betrieb einer höheren Strahlenbelastung, als nach der Strahlenschutzverordnung zulässig, ausgesetzt würden.

(SVA-Bulletin)

BRD. Auf Empfehlung der Enquête-Kommission „Zukünftige Kernenergie-Politik“ des Deutschen Bundestages hat die Gesellschaft für Reaktorsicherheit (GRS) eine risikoorientierte Analyse zum Prototyp-Schnellbrüter-Kernkraftwerk SNR-300 durchgeführt, um einen Sicherheitsvergleich zu einem Druckwasserreaktor vom Typ Biblis zu ermöglichen. Über die Studie berichtet Köberlein Garching von der GRS in der „atomwirtschaft-atomtechnik“ (Nr. 8/9, August/September 1982 — S. 429 — 435). Danach zeigen die Ergebnisse der Analyse, daß für den SNR-3000 am Standort Kalkar sowohl die theoretische Häufigkeit schwerer Unfälle als auch das dafür berechnete Schadensausmaß geringer sind als bei einem Druckwasserreaktor, wie sie in der deutschen Risikostudie für Leichtwasserreaktor-Kernkraftwerke ermittelt wurden.

Wegen des niedrigen Drucks, des hohen Siedepunktes von 950 °C und der hohen Wärmekapazität des Kühlmittels Natrium spielen Kühlmittelverluststörfälle und Ausfälle der Nachwärmeabfuhrsysteme nur eine untergeordnete Rolle. Das größte Risiko für Schäden am Reaktorkern besteht bei gleichzeitigem Ausfall der Kühlmittelpumpen und der Abschalteinrichtungen. In einer solchen Situation kann es zum Sieden des Natriums und damit zu einer Vermehrung der Kettenreaktion kommen. Auf Grund der mehrfachen Sicherheitsbarrieren des SNR-300 wird eine weitgehende Rückhaltung radioaktiven Materials auch bei einer — sehr unwahrscheinlichen — Zerstörung des Reaktorkerns erreicht. Die Wahrscheinlichkeit der Zerstörung des Reaktorkerns des SNR-300 wird mit 1 in 500000 Jahren angegeben. Demgegenüber nennt die vor drei Jahren fertiggestellte Risikostudie eine Wahrscheinlichkeit für die Zerstörung eines Leichtwasserreaktor-Kerns von 1 in 10000 Jahren. (P. Z.)

(SVA-Bulletin)

Schweiz. In der Zeit vom 23.—24. Juni 1982 wurde durch das Bundesamt für Gesundheitswesen eine Sammlung von schwach radioaktiven Abfällen aus Spitälern, Hochschulen und Forschungsinstituten sowie Radioisotopen verarbeitenden Betrieben durchgeführt. Insgesamt 184 Stahlfässer zu 100 l, 16 Stahlfässer zu 200 l sowie 11 Kleinbehälter aus 32 Betrieben wurden dem Eidgenössischen Institut für Reaktorforschung in Würenlingen zur Sortierung, Konditionierung und anschließender Meerversenkung zugeführt. Der Inhalt dieser Fässer, bestehend aus festen und verfestigten radioaktiven Abfällen, vermischt mit verschiedenartigen anderen festen Abfallstoffen, wies Aktivitäten zwischen 10 µCi bis 1000 Ci pro Faß auf.

(SVA-Bulletin)

Schweiz. Nach sieben Gemeinden im Kanton Aargau hat nun auch die Zürcher Gemeinde Weiach der Nagra (Nationale Genossenschaft für die Lagerung radioaktiver Abfälle) die Baubewilligung für eine Probebohrung auf ihrem Gebiet erteilt. Damit fehlen der Nagra für die ersten zwölf geplanten Probebohrungen in der Nordschweiz nur noch vier Bewilligungen. Weiach wird voraussichtlich neben dem aargauischen Böttstein die zweite Gemeinde sein, wo die Nagra diesen Herbst mit dem Bohren beginnt. In Böttstein arbeitet man bereits seit Juli an der Herrichtung des Bohrplatzes. Während der Nagra-Bohrungen soll ein umfangreiches Forschungsprogramm durchgeführt werden. In einem Gebiet von 1200 Quadratkilometern untersucht die Nagra die Eignung des Grundgebirges aus Granit und Gneis für die Endlagerung der stark radioaktiven Abfälle.

(SVA-Bulletin)

Schweiz. Die Schweiz beteiligt sich zur Zeit erneut an einer Meeresversenkungsaktion für leicht radioaktive Abfälle. Wie in früheren Jahren wird diese gemeinsam mit Holland und Belgien durchgeführt. Die Kernkraftwerke, die im Jahre 1981 rund 14,4 Mia kWh elektrische Energie erzeugten, liefern rund 60% des Abfallvolumens, der Rest stammt aus Medizin, Forschung und Industrie. Es handelt sich um Filter, Schlämme, Ionenaustauscherharze, Asche aus der Verbrennungsanlage sowie verschiedene kontaminierte Materialien. Die eigentlich strahlenden Atome des gesamten Abfalls machen weniger als 1 kg aus. Sie sind mit 850 Tonnen inaktivem Beton in 880 Fässern und 152 Betoncontainern eingegossen. Pro kWh erzeugte elektrische Energie entsteht so weniger als 1/10 Gramm schwachaktiver Abfall inklusive Beton. Die Aktivität aus den Kernkraftwerken ist zudem kurzlebig und zerfällt lange bevor die Behälter durch Korrosion Schaden nehmen können. Der größte Teil der Aktivität kommt übrigens nicht von den Kernkraftwerken, sondern stammt aus den Tritiumabfällen der Leuchtfarbenindustrie.

Die Verpackung erfolgte aufgrund von internationalen Richtlinien, und die ganze Ladung wurde von den eidgenössischen Sicherheitsbehörden für die Versenkung freigegeben. Der Transport nach Belgien in einem Sonderzug mit verschlossenen Bahnwagen erfüllt die hohen Anforderungen der internationalen Transportvorschriften. Für die Versenkung gilt die von der Schweiz ratifizierte Londoner Konvention über die Verhinderung der Meeresverschmutzung. Der Lagerplatz im Atlantik befindet sich in 4500 m Tiefe ungefähr 900 km südwestlich der Südspitze Englands und ist der gleiche wie in früheren Jahren.

(SVA-Bulletin)

Frankreich. In der Versuchsanlage SAP (Services de l'Atelier Pilote) sind seit 1974 7,5 t verschiedener oxidischer Kernbrennstoffe, darunter 3,0 t aus dem Schnellbrüter-Kraftwerk Phénix wiederaufgearbeitet worden. Die Anlage SAP, die schon 1962 in Betrieb ging, wird jetzt durch das Projekt TOR (Traitmente des Oxydes Rapides) ergänzt und erweitert, um in Marcoule, wo das Commissariat à l'Energie Atomique die Forschungs- und Entwicklungsarbeiten für die Wiederaufarbeitung konzentriert, über eine moderne integrale Versuchsanlage für Schnellbrüter-Brennstoffe verfügen zu können. Die Ergänzungen betreffen insbesondere Einrichtungen für den mechanischen Aufschluß und für die Auflösung der bestrahlten Brennelemente. Daneben wird die Kapazität auf 5 t Uran pro Jahr erhöht. Das Projekt sieht die Installation zweier paralleler Linien vor, von denen die eine zur Schließeung des Brennstoffkreislaufes der Schnellbrüterreaktoren Rapsodie und Phénix dient. Die zweite Linie hat ausschließlich experimentellen Charakter und wird für die Erprobung von neuen, fortschrittlichen Apparaten eingesetzt werden. Die um TOR erweiterte Versuchsanlage dient der Vorbereitung auf den Bau der Prototypanlage PURR (Prototype d'Usine de Retraitement des Rapides), deren Inbetriebnahme für den Beginn der 90er Jahre geplant ist. Wie bei den neuen Anlagen in La Hague ist auch bei TOR die Société Générale pour les Techniques Nouvelles (SGN) mit den Ingenieurarbeiten betraut. Mit dem Bau von TOR wurde 1980 begonnen.

(SVA-Bulletin)

Großbritannien. Drei Gesellschaften mit Erfahrung in der Behandlung radioaktiver und toxischer Materialien haben sich in Großbritannien zusammengeschlossen, um Ausrüstungen und Dienstleistungen auf dem Gebiet der Abfallbehandlung anzubieten. Die von ihnen gegründete Gesellschaft unter dem Namen NEI Waste Technologies wird zu 55% von der Northern Engineering Industries, zu 35% von der amerikanischen Chem Nuclear Systems und zu 10% von der Pollution Prevention Consultants of Crawley gehalten. Durch die Aufforderung der britischen Regierung an die Privatwirtschaft zur Betätigung auf diesem Gebiet ermutigt, sieht die neue Gesellschaft ihr Betätigungsfeld in der Behandlung und dem Transport schwach- und mittelaktiver Abfälle.

(SVA-Bulletin)

USA. Wie jetzt mit Hilfe einer kleinen Unterwasser-Fernsehkamera im Reaktorkern des amerikanischen Kernkraftwerksblocks Three-Mile Island-2 (TMI-2) bestätigt werden konnte, sind beim Störfall vom 28. März 1979 große Teile der Brennelemente zerbröckelt und die Bruchstücke im Reaktorkern nach unten gefallen. Infolge Überhitzung des Reaktorkerns oxydierten die Brennstabhüllen aus Zirkaloy, und die Uranbrennstoff-Tabletten fielen nach unten. Dagegen ergaben die TV-Bilder entgegen Presseberichten keine Hinweise darauf, daß Brennstofftabletten geschmolzen wären.

Die TV-Inspektion wurde am 21. Juli 1982 durchgeführt. Der anderthalbstündige Videofilm wird noch im Detail ausgewertet. Vorläufig wurde festgestellt, daß der visuelle Befund keine Überraschungen brachte und die Ergebnisse verschiedener theoretischer Studien über den mutmaßlichen Zustand des Reaktorkerns und der Reaktoreinbauten zu bestätigen scheint. An der inspizierten Position waren an den Reaktoreinbauten oberhalb der Brennelemente visuell keine Schäden feststellbar. Von den Brennelementen war längs der oberen anderthalb Meter nichts mehr zu sehen. Unten traf die Kamera dann tatsächlich ein Schuttbett herabgefallener Brennstofftrümmer.

Vor der TV-Inspektion hatte die GPU Nuclear Corporation am 13. Juli mitgeteilt, die Reinigung des radioaktiv kontaminierten Reaktorkühlwassers von TMI-2, die noch mehrere Monate weiterlaufen wird, habe bereits eine beträchtliche Reduktion der Kühlwasser-Aktivität gezeigt. Für Caesium-134 sei ein Rückgang der Aktivitätskonzentration von 1,4 auf 0,2 µCi/ml gemessen worden und für Caesium-137 ein solcher von 13,5 auf 2,4 µCi/ml. Bei Strontium-90 habe man eine Reduktion von 16,0 auf 9,3 µCi/ml erreicht.

(SVA-Bulletin)

Japan. In Japan gelang kürzlich die Weltpremiere, Uran in Gramm-Mengen aus Meerwasser zu gewinnen. Wie das Government Industrial Research Institute in Shikoku am 10. Juni 1982 mitteilte, wurden in einer Versuchsanlage 4,1 g Urankonzentrat aus natürlichem Meerwasser extrahiert. An den Arbeiten sind neben dem Institut in Shikoku, das dem Ministry of International Trade and Industry (MITI) untersteht, das Institute of Industrial Science der Universität Tokyo, die Mitsubishi Chemical Industries Ltd., sowie die Unitika, Ltd., beteiligt. Das angewandte Verfahren besteht in der Absorption des im Meerwasser gelösten Urans an einem körnigen Absorbermaterial und der anschließenden Herauslösung sowie Aufkonzentration des adsorbierten Urans. Die Regierung Japans, das über keine ins Gewicht fallenden konventionellen Uranvorkommen verfügt, unterstützt die Forschungsarbeiten zur Urangewinnung aus dem Meer, obwohl 1 m^3 Meerwasser nur rund 3 mg Uran enthält.

(SVA-Bulletin)

Japan. Der Electric Power Resources Development Coordination Council (EPRDCC) der japanischen Regierung genehmigte am 9. Juli den Elektrizitätsausbauplan für das Finanzjahr 1982. Danach sollen 1982 die Bauarbeiten für Elektrizitätswerke mit einer Gesamtleistung von 10000 MW aufgenommen werden, wovon 5000 MW auf konventionell-thermische Anlagen, 4000 MW auf Kernkraftwerke und 1000 MW auf Wasserkraftwerke entfallen. Für die Achtjahresperiode 1982 bis 1989 rechnet der Plan mit der Stilllegung von veralteten Kraftwerken im Umfang von 2600 MW und mit der Inbetriebsetzung neuer Elektrizitätswerke mit einer Leistung von 69 700 MW. Der Atomstromanteil soll bis zum Ende des Finanzjahres 1989 auf 17% (Ende 1981: 12%) angehoben werden, während dann 29% (1981: 46%) des japanischen Stroms aus Öl, 22% (16%) aus Naturgas, 11% (4%) aus Kohle und 21% (22%) in Wasserkraftwerken produziert werden soll. Gemäß der Planung des Ministry of International Trade and Industry (MITI) sollen dem EPRDCC im Finanzjahr 1982 die Baugesuche für die vier Nuklearblöcke Higashidori-1 (1100 MW, BWR), Ikata-3 (890 MW, PWR) sowie Genkai-3 und -4 (je 1180 MW, PWR) unterbreitet werden.

(SVA-Bulletin)

IAEA. Anfang 1977 erstellte die IAEA ein neues Programm über Standardisierung und Vergleich hoher Strahlendosen bei industriellen Strahlenverfahren. Ausschlaggebend dafür war die Beobachtung, daß die Verwendung hoher Strahlendosen zur Sterilisierung medizinischer Präparate, zur Behandlung von Klärschlamm und zur Bearbeitung verschiedener heute weitverbreiteter Kunststoffe im großen Maßstab üblich geworden ist. Bei der Bestrahlung von Lebensmitteln wurden wegen deren Genießbarkeit und Aufnahme durch die Öffentlichkeit besonders strenge Anforderungen in bezug auf die genaue Dosimetrie gestellt. So erhielt die Messung hoher Strahlendosen, die bisher eher vernachlässigt worden war, eine wichtige Funktion in der industriellen Qualitätskontrolle und erregte das Interesse sowohl der Techniker als auch der Manager. Dieses internationale Programm und die in seinem Rahmen gebotenen Dienstleistungen werden als einzigartig anerkannt. Bisher hat keine internationale Stelle, die im Bereich der Strahlendosimetrie tätig ist, ein Standardisierungsprogramm für Messungen derartiger Dosen eingeleitet.

Seit 1978 läuft das koordinierte Forschungsprogramm, in dessen Rahmen elf Wissenschaftler Vergleiche von Messungen hoher Strahlendosen durchführen. Ziel dieser Forschungsarbeiten ist die Lösung der bei Dosimetrievergleichen auftretenden Probleme und die Verbesserung der Meßverfahren bzw. die Entwicklung neuer Dosimetriesysteme. Das Seminar über die Messung hoher Strahlendosen bei industriellen Strahlenverfahren, das die IAEA vom 20. September bis 1. Oktober 1982 in Roskilde, Dänemark, veranstaltete, ist ebenso wie ein internationales Symposion über das Thema, das 1984 stattfinden soll, ein weiterer Schritt in dieser Richtung.

Letztes Ziel des Programms ist die Schaffung einer internationalen Stelle für Dosisvereinheitlichung, die den Mitgliedstaaten bei verschiedenen Projekten unter Verwendung hoher Strahlendosen an die Hand gehen soll. Damit werden Empfehlungen der gemeinsam von der Ernährungs- und Landwirtschaftsorganisation der Vereinten Nationen (FAO) und der Weltgesundheitsorganisation (WHO) unterhaltenen Codex-Alimentarius-Kommission befolgt, die die Notwendigkeit entsprechender dosimetrischer Aufzeichnungen an Anlagen für Lebensmittelbestrahlung unterstreichen. Auch der 1980 tagende gemeinsame Expertenausschuß der FAO, IAEA und WHO über die Genießbarkeit bestrahlter Lebensmittel betonte, daß eine entsprechende Kontrolle der Strahlendosen eine Voraussetzung für die praktische Anwendung der Nahrungsmittelbestrahlung sei. Der Ausschuß stellte auch fest, daß die IAEA im Rahmen ihres Programms über die Messung hoher Strahlendosen Unterstützung im Bereich der Dosiskontrolle bietet.

Das gegenwärtige Seminar ist die erste einschlägige Tagung, die die IAEA veranstaltet. Es soll eine gründliche Spezialausbildung in Dosimetrieverfahren vermitteln, die für den Betrieb von Bestrahlungsanlagen mit Kobalt-60 oder Elektronenstrahlen benötigt werden. Das wissenschaftliche Programm umfaßt Vorlesungen, experimentelle Arbeiten im Laboratorium, sowie Diskussionen über einschlägige Themen. Bei Versuchen im Labor werden die Tagungsteilnehmer Gelegenheit haben, sich mit den modernsten und zuverlässigsten routinemäßigen Dosimetriesystemen vertraut zu machen.

(IAEA-Pressemitteilung)

IAEA. Das beste Geburtstagsgeschenk für die IAEA, die ihr 25jähriges Bestehen feierte — und die höchste Anerkennung für die Idealisten, die an ihrer Gründung beteiligt waren, sie über die Anfangsschwierigkeiten hinweggebracht und die Begeisterung für ihre Ziele nie verloren haben —, wäre es, ihren Ruf als technische, objektive Organisation zu erhalten, Geduld und Zurückhaltung zu üben, wenn die Vertreter der Mitgliedstaaten heikle Fragen zur Sprache bringen, und sich auf jene Bereiche zu konzentrieren, wo die Organisation das meiste leisten kann. Diese Anregung und diesen verhüllten Appell richtete der Generaldirektor der IAEA, Dr. *Hans Blix*, in seiner jährlichen Ansprache an die 26. ordentliche Tagung der Generalkonferenz. Er wies darauf hin, daß unmittelbar vor der Generalkonferenz eine große internationale Konferenz über Erfahrungen mit der Kernenergiegewinnung stattgefunden hat, auf der Gelegenheit zu einer Bestandsaufnahme gewesen sei. „Der überschäumende Optimismus der Sechzigerjahre ist durch die Schwierigkeiten der letzten Jahre gedämpft worden", sagte er, „In einigen Ländern stagniert die Kernenergie, in anderen hat sich ihr Wachstum verlangsamt und bei den Programmen sind Verzögerungen eingetreten. Einige Länder führen ihre ursprünglichen Pläne weiter aus, aber nur in wenigen gibt es Programme mit rascher Expansion. „Die Erkenntnis der gegenwärtigen Schwierigkeiten darf einen aber nicht zu dem übereilten Schluß veranlassen, daß die Kernenergie keine Zukunft hat", meinte er. „Denn die Zukunft von gestern, ein Gebiet, das vor 25 Jahren auf Versuche im Laboratorium beschränkt war, ist jetzt in vielen Ländern eine Industrie großen Maßstabes. 9% der Stromproduktion der Welt stammen heute aus Kernkraftwerken, und voraussichtlich wird die Kernenergie bis zum Ende dieses Jahrzehnts 20% der auf der Welt erzeugten Elektrizität beisteuern. Darüber hinaus müssen wir an die Zukunft denken, wenn die Entwicklungsländer einen größeren Teil des auf der Welt produzierten Erdöls für sich beanspruchen müssen. Dann werden wir die Kernenergie und andere Energieträger brauchen, um die Erdölproduktion zu entlasten."

Dr. *Blix* wies auf einige Aufgaben hin, mit denen die IAEA konfrontiert ist:

— Erstens müssen die Anlauf- und Bauzeiten für Kernkraftwerke verkürzt werden, um die Kosten herabzusetzen, ohne aber die hohen Anforderungen an Sicherheit zu beeinträchtigen. Der beste Weg dazu ist die Standardisierung der Anlagen und der wichtigsten Anlageteile, die im innerstaatlichen und sogar im internationalen Maßstab von den Herstellern, Energieversorgungsunternehmen und Aufsichtsbehörden gemeinsam gefördert werden muß. Es wäre nützlich, international vereinbarte Grundkriterien für Sicherheit, Erprobung und Qualitätsstandardisierung, Industrienormen und eine rationelle Durchführung von Projekten auszuarbeiten.

— Zweitens könnte die durchschnittliche Verfügbarkeit der Kernkraftwerke verbessert werden. Die Anlagen mehrerer Länder haben klar bewiesen, daß eine Verfügbarkeit von nahezu 80% und manchmal mehr erreichbar ist. Der durchschnittliche Lastfaktor der arbeitenden Anlagen liegt bei 62% — was einen bedeutenden Verlust an potentieller Stromproduktion aus Kernenergie darstellt. Ein intensiver internationaler Informationsaustausch, unter anderem die Beistellung von Daten für das Dokumentationssystem der IAEA für Leistungsreaktoren (PRIS) und die Auswertung dieser Daten könnte dazu beitragen, die Verfügbarkeit ohne Einbuße an Sicherheit zu verbessern.

— Drittens brauchen Entwicklungsländer, die ein Kernenergieprogramm in Angriff nehmen wollen, Unterstützung. Die IAEA kann sie bei der Wahl geeigneter Methoden beraten, um die Vorteile der Kernenergie gegenüber jenen anderer Energieträger abzuwägen. Sobald die Inangriffnahme eines Kernenergieprogramms beschlossen ist, kann die IAEA durch Ratschläge beim Aufbau der nötigen Infrastruktur und der Ausbildung von Personal behilflich sein.

Die Forschung an den Grenzen der Technik muß weitergehen, um für die nächste Generation von Reaktoren den Weg zu bereiten. Reaktoren fortschrittlicher Bauart könnten eine intensivere Nutzung der Weltvorräte an Kernbrennstoffen ermöglichen. Dr. *Blix* wies darauf hin, daß die IAEA, um einen breiteren Gedankenaustausch über dieses Thema zu fördern, im Begriff ist, eine Untersuchung über den potentiellen Beitrag solcher Reaktoren zur Energieversorgung der Welt zu beginnen. Wenigstens in Form einer Vorstudie sollen dabei auch Leistungsreaktoren späterer Generationen und die Kernverschmelzung einbezogen werden, letztere ein Gebiet, auf dem die IAEA entscheidende Schritte unternommen hat — zum Beispiel mit der Studie über einen internationalen Tokamak-Reaktor (INTOR). Ebenfalls behandelt werden kleinere Reaktoren, die bei der Stromerzeugung in Entwicklungsländern und auf neuen Anwendungsgebieten, vor allem für Fernheizungszwecke, eine wichtige Rolle spielen könnten.

Zum Problem nukleare Sicherheit

Ein Beweis für die zunehmende Anteilnahme der IAEA an Fragen der nuklearen Sicherheit sei die Veröffentlichung eines Allgemeinen Berichts über nukleare Sicherheit für 1980—81 im Juni gewesen, sagte Dr. *Blix*. Es sei geplant, jedes Jahr einen solchen Bericht herauszubringen, um dazu beizutragen, daß auf diesem Gebiet weniger lückenhafte und zuverlässigere Informationen vorliegen.

Durchschnittlich seien die heutigen Kernkraftwerke seit 12 Jahren in Betrieb, meinte Dr. *Blix*, einige seien es schon seit 25 Jahren. „Aus diesen Erfahrungen können wir mehrere Lehren ziehen,“ sagte er: „die erste und erfreulichste ist, daß die Kernkraftwerke sicher gearbeitet haben. Es hat keinen Unfall mit Todesopfern gegeben, die direkt durch Strahlen verursacht worden sind. Zweitens haben selbst bei einem schweren Unfall, wie dem von Three Mile Island, die nuklearen Sicherheitssysteme funktioniert. Selbst bei einem Zusammenwirken von menschlichem Fehlverhalten und mechanischen Ausfällen ist die freigesetzte Radioaktivität im wesentlichen eingeschlossen geblieben. Aus dem Unfall konnten aber die Kernindustrie und die Anlagebetreiber in aller Welt Wesentliches lernen, und eine unmittelbare Folge davon waren Verbesserungen der Sicherheit der Anlagen.“

Ein rascher Abschluß der Wiederherstellungsarbeiten in Three Mile Island würde die Besorgnisse der Öffentlichkeit mildern. Und, was noch wichtiger ist, wenn in der gesamten Industrie der bisherige hohe Stand der Betriebssicherheit aufrechterhalten wird, so würde man der Kernenergie positiver gegenüberstehen.

Auch bei Forschungen über Sicherheitsfragen sei eine Zusammenarbeit sehr nützlich. Die bestehenden Vereinbarungen könnten durch verstärktes Zusammenwirken verschiedener regionaler Gruppen erweitert werden. Die Erstellung von Alarmplänen sei ebenfalls ein Gebiet, wo innerhalb der IAEA neue Initiativen für internationale Zusammenarbeit erörtert würden.

Zum Problem Entsorgung

Dr. *Blix* wies darauf hin, daß fossile Brennstoffe und die aus ihnen entstehenden Abfälle wichtige Schadstoffe sind. Die Besorgnisse der Öffentlichkeit wegen der radioaktiven Abfälle — deren Volumen gering sei — habe die günstige Wirkung gehabt, die Forschung zu intensivieren und qualitativ hochwertige Ergebnisse zu zeitigen. „Die Kernwissenschaftler und -techniker sind sich allgemein darin einig, daß es heute schon zufriedenstellende technische Lösungen gibt“, sagte Dr. *Blix* und äußerte die Hoffnung, daß die vor kurzem von der IAEA und der Weltgesundheitsorganisation herausgebrachte Broschüre „Kernenergie, Umwelt und Mensch“ von der Öffentlichkeit und den Medien als unvoreingenommener und sachlicher Diskussionsbeitrag aufgenommen werden wird.

Die Wiederaufarbeitung scheine sich als beliebteste Methode für den Umgang mit abgebrannten Kernbrennstoffen zu erweisen, da sie das Problem der Zwischenlagerung mildere und es ermögliche, hochaktive Abfälle in eine ungefährlichere Form überzuführen, in der sie leichter zu manipulieren sind, und gleichzeitig Plutonium als Brennstoff für Brüter zu erzeugen. Bis aber für die Wiederaufarbeitung eine klare politische Linie gefunden und eine ausreichende Kapazität aufgebaut sei, werde sich weiter abgebrannter Brennstoff ansammeln, und der dringende Bedarf an zusätzlicher Lagerkapazität werde weiter bestehen bleiben.

Zur Frage der Zwischenlagerung abgebrannter Brennstoffe bringe der eben erschienene Bericht einer Expertengruppe „sehr praktische“ Anregungen in bezug auf technische Hilfe, Informationsaustausch, Richtlinien und Normen. „Wenn die für abgebrannte Brennstoffe verantwortliche Industrie und die Mitgliedstaaten die Ideen der Experten über eine nationale oder internationale Zusammenarbeit bei der Lagerung abgebrannter Brennstoffe erwägenswert finden, werden sie vielleicht bereit sein, praktische Vereinbarungen darüber auszuarbeiten,“ meinte Dr. *Blix*. „Die IAEA kann, wenn es gewünscht wird, den Rahmen für die nötigen Kontakte stellen.“

Zum Problem Handel auf nuklearem Gebiet

Dr. *Blix* stellte fest, daß eine internationale Einigung über den Handel auf kerntechnischem Gebiet noch aussteht, und äußerte die Hoffnung, daß der Ausschuß der IAEA für Versorgungsgarantien imstande sein werde, einen Ausweg aus der gegenwärtigen Sackgasse zu finden. Das Hauptproblem, das Anlaß für die Veranstaltung der Konferenz der Vereinten Nationen über die „Förderung der Internationalen Zusammenarbeit bei der Friedlichen Nutzung der Kernenergie" gewesen sei, die 1983 in Genf stattfindet, bestehe zweifellos darin, eine ausreichende Versorgung gleichzeitig mit einem entsprechenden Schutz gegen eine Weiterverbreitung von Kernwaffen zu gewährleisten — eine Frage, auf die der Ausschuß für Versorgungsgarantien noch keine vollständige Antwort gefunden habe. Länder, die nur über kleine Programme verfügen, brauchten echte Garantien für alle Phasen des Brennstoffkreislaufs, einschließlich der Entsorgung.

Zum Problem Sicherheitskontrolle

„Immer mehr setzt sich die Überzeugung durch, daß eine nukleare Abrüstung dringend nötig ist", sagte Dr. *Blix*. „Man ist sich auch bewußt, daß die Welt noch gefährlicher würde als sie es heute ist, wenn es zu einer weiteren Ausbreitung der Kernwaffen käme". Die IAEA sei fest entschlossen, alle dagegen gerichteten Bestrebungen zu unterstützen. „Eine wesentliche Grundlage für alle Maßnahmen gegen die Ausbreitung der Kernwaffen ist die ständige politische Entschlossenheit der Staaten, aber das Überwachungssystem der IAEA leistet einen sehr wichtigen Beitrag, da es überprüft, ob die eingegangenen Verpflichtungen eingehalten werden, und dadurch das nötige Vertrauen schafft."

Die Überwachungstätigkeit sei nicht frei von Schwierigkeiten, „aber die ernsteren Probleme, die außerhalb der Kompetenz der IAEA liegen, beginnen — das behaupte ich — dort, wo die Sicherheitskontrolle endet", sagte er.

Dr. *Blix* gab zu, daß die IAEA durch ihre Überwachungstätigkeit nichts über die künftige Politik der Staaten erfährt und nichts durch physisches Eingreifen verhindern kann; sie kann nur die gegenwärtigen Arbeiten überprüfen und darüber Bericht erstatten. „Trotzdem ist die Sicherheitskontrolle etwas Einzigartiges. Zum ersten Mal in der Geschichte haben souveräne Staaten eine internationale Organisation aufgefordert, wichtige Anlagen auf ihrem Territorium zu inspizieren. Das muß mit Takt geschehen, aber auch mit der Gründlichkeit, die die nötige Gewähr gibt". Die IAEA selbst sei die erste, die die Mängel des gegenwärtigen Überwachungssystems untersucht und Verbesserungen vorschlägt. Sie brauche mehr Arbeitskräfte und mehr Ausrüstungen, um mit dem wachsenden Arbeitsanfall fertig zu werden. Ebenso habe sie erkannt, daß die Organisation im Zuge ihres Wachstums so umgestaltet werden muß, daß sie ihren Aufgaben möglichst gut gerecht werden kann.

Da aber die Möglichkeit einer Abzweigung durch einen Staat, nicht durch einen Anlagebetreiber, Gegenstand der Sicherheitskontrolle ist, wäre es vielleicht angebracht, den Zusammenhängen zwischen allen Anlagen des Brennstoffkreislaufs in einem Staat stärkeres Augenmerk zu schenken. Vielleicht wird die Ständige Beratergruppe über Durchführung der Sicherheitskontrolle (SAGSI) Vorschläge in dieser Richtung machen. Ungelöst ist auch noch die Frage, wie große Wiederaufarbeitungs- und Anreicherungsanlagen am besten zu überwachen sind. In kernwaffenlosen Staaten wird es noch mehrere Jahre lang keine solchen Anlagen geben, wir haben also Zeit, dieses Problem zu lösen; vielleicht wäre es eben am besten, uns schon jetzt damit auseinanderzusetzen, wo diese Anlagen geplant werden."

Dr. *Blix* erwähnte, es werde manchmal die Frage gestellt, ob das Überwachungssystem der IAEA in dem größeren Bereich der Waffenkontrolle als Vorbild dienen könnte. „Die Sicherheitskontrolle erstreckt sich auf friedliche Anlagen — nicht auf Waffen —, und sie beschränkt sich auf bestimmte Anlagen und Materialien. Das sind Einschränkungen. Trotzdem aber ist sie ein wichtiger Beweis dafür, daß eine Überprüfung an Ort und Stelle eine politisch tragbare und mögliche Methode ist, Vertrauen zu erzeugen".

(IAEA-Pressekommuniqué)

Großbritannien. Die britische Atomenergie-Agentur (UK AEA) hat folgende Berichte herausgegeben:

AERE-M 3235	Analysis of Zirconium alloys using inductively-coupled plasma emission spectrometry. By *G. F. White* and *C. J. Pickford*. March 1982. 11 pp.
AERE- 9887	The uptake of radionuclides by plants: A review of recent literature. By *P. A. Cawse* and *G. S. Turner*. February, 1982. 47 pp.
AERE-R 10306	The calculation of spectra and effective energies of β-decay. By *L. J. Baker*. May, 1982. 25 pp.
AERE-R 10402	Some Monte-Carlo calculations of resolution effects in neutron detectors for resonance neutron radiography. By *G. Robertson*. April, 1982. 16 pp.
AERE-R 10429	= TF — A library of TSO facilities. By *D. A. Lever* and *P. C. Robinson*. April, 1982. 57 pp.
AERE-R 10432	MORSE-H: A revised version of the Monte Carlo code MORSE. By *N. P. Taylor* and *J. Needham*. May, 1982, 32 pp.
AERE-R 10521	The characterisation of precipitated magnetites. By *D. F. Rush* and *D. L. Segal*. June, 1982. 25 pp.
CLM-R 216	Atomic and molecular data for fusion, Part 1. Recommended cross sections and rates for electron ionisation of light atoms and ions. By *K. L. Bell*, *H. B. Gilbody*, *J. G. Hughes*, *A. E. Kingston* and *F. J. Smith*. December, 1981. 120 pp.
CLM-R 220	Simple analytic form of the relativistic Thomson scattering spectrum. By *A. C. Selden*. January, 1982. 21 pp.
CLM-R 224	The toroidal field coil structure for the Culham Mark IIC Tokamak reactor. By *S. F. Calvert*, *D. H. Prothero* and *W. R. Spears*. 1982. 61 pp.

(Atom 311)

Dänemark. Anläßlich der 23. ordentlichen Generalversammlung des Schweizerischen Vereins für Atomenergie (SVA) am 19. August 1982 in Bern hielt Ing. *Søren Mehlsen*, Leiter der Nuklearabteilung der ELSAM Kraftvaerksafdelingen, Fredericia, und Präsident der Dänischen Kerntechnischen Gesellschaft, einen Vortrag über „Dänemark — ein Land ohne Kernenergie". *S. Mehlsen* analysiert zunächst die Energieversorgung Dänemarks im Zeitraum 1970—2000 und führt u. a. aus: Während der 60er Jahre wurde der dänische Energiemarkt völlig von den Erdölerzeugnissen beherrscht. Heizöl gewann mehr und mehr Verbreitung im Raumheizungssektor, während Dieselkraftstoff und Benzin als Treibstoffe den Transportsektor dominierten und Industrieanlagen sowie Kraftwerke für den Schweröleinsatz ausgelegt wurden. Selbstverständlich wurden alle Energierohstoffe auf dem internationalen Markt eingekauft, der meistens ein Käufermarkt war.

Bekanntlich änderte sich dies 1973 drastisch. Das OPEC Kartell setzte die ersten massiven Aufschläge auf den Erdölpreisen durch, und die Handelsbilanz der Industrieländer, die bereits von einer Rezession gezeichnet war, kam zusätzlich durch Energieimporte unter Druck.

Die Antwort der dänischen Regierung und des Parlaments, des Folketings, bestand in der Einleitung einer Reihe energiepolitischer Maßnahmen. Sie verfolgten damit drei Hauptziele:

- Erstens sollte generell beim Energieverbrauch und besonders bei der Raumheizung gespart werden, um den Druck auf die Handelsbilanz zu reduzieren.
- Zweitens sollten vermehrt andere Energierohstoffe, besonders Kohle, eingesetzt werden, um einerseits den Bedarf an importiertem Erdöl zu verringern und andererseits die Versorgungssicherheit durch eine breitere Abstützung der Einfuhren zu verbessern.
- Drittens sollten die Öl- und Erdgasfelder im dänischen Teil der Nordsee erschlossen werden.

Dänemark verschrieb sich also, wie die anderen Industrieländer, dem Sparen, dem Substituieren und dem Diversifizieren.

Die Energiepolitik der dänischen Regierung zielt jedoch auf eine breitere Abstützung der Energiequellen im Laufe der 90er Jahre. Dies wäre durch die Einführung der Kernenergie zu erreichen. In der Importstatistik würde in diesem Fall ein Teil der Kohle durch Uran substituiert.

Der Bau von Kernkraftwerken ist indes an mehrere Voraussetzungen geknüpft. Dazu gehört eine befriedigende Beantwortung von Fragen nach der Endlagerung radioaktiver Abfälle und nach den Auswirkungen großer Störfälle. Beide werden noch bis Anfang 1983 von den Umweltschutzbehörden studiert. Zur Erdgasversorgung wird festgestellt, daß die seit 1963 erforschten KW-Vorkommen im dänischen Sektor der Nordsee etwa 200 km westlich von Jütland liegen. Bei einer jährlichen Förderung von vier Millionen Tonnen Rohöl und 2—3 Milliarden Kubikmeter Erdgas ab Mitte der 80er Jahre dürften die nachgewiesenen Ressourcen bis ins erste Jahrzehnt des kommenden Jahrhunderts reichen.

Die staatliche Aktiengesellschaft DONG (Dansk Olie og Natur Gas) besitzt die Exklusivrechte für die Einfuhr, den Transport, die Lagerung und den Verkauf von Erdgas in Dänemark. DONG kann neben den einheimischen Vorräten auf Lieferungen der deutschen Ruhrgas AG zurückgreifen. Dies ermöglicht ihr jetzt schon den schrittweisen Ausbau eines Versorgungsnetzes, bis ab etwa 1984 die Lieferungen aus den Nordseefeldern beginnen können.

Und nun die Feststellungen *S. Mehlsens* zur Kernenergie: Neben den großen Anstrengungen zur Sicherung der Elektrizitäts- und Fernwärmeversorgung durch die Umstellung auf Kohlefeuerung hat die dänische Elektrizitätswirtschaft auch Nuklearprojekte vorbereitet. Vor dem Ölpreisschock von 1973 hatte sie allerdings nur geringes Vertrauen in die Aussichten der Kernenergie. Angesichts der Situation auf dem Schwerölmarkt, der als Käufermarkt mit fallender Preistendenz beurteilt wurde, errechnete man als Rentabilitätsschwelle für eine nukleare 600 Megawatt-Einheit einen Ölpreis von 4 Kronen je Gigajoule, und der damalige Preis lag bei nur 2 Kronen. Trotzdem nahmen die Vorbereitungen für die Einführung der Kernkraft in Dänemark langsam Gestalt an; ab 1974 wurden sie konkreter, dann jedoch durch die allgemeinen Entwicklungen auf dem Energiegebiet überrollt. Sie veranlaßten den Staat, die Führungsrolle zu übernehmen. So brachten es die größeren Kompetenzen der Regierung auch mit sich, daß ein Entscheid wie die Einführung der Kernenergie der Genehmigung durch Regierung und Parlament unterworfen wurde.

Zuerst muß das Umweltministerium einen Bericht der Elektrizitätswerke über die Möglichkeit zur sicheren Lagerung hochradioaktiver Abfälle gutheißen. Zur Betonung ihres ernsthaften Interesses an der Kernenergie haben Elkraft und Elsam 1979 bis '80 gemeinsam mit einem Aufwand von 80 Millionen Dänenkronen umfassende geologische Abklärungen getroffen, so mehrere kilometertiefe Sondierbohrungen in einen Salzdom in Jütland abgeteuft und einen Sicherheitsbericht ausgearbeitet. Das Umweltministerium hat sich eine Prüfungsfrist bis zum Frühling des kommenden Jahres gesetzt. Gleichzeitig erarbeitet es unter Beizug in- und ausländischer Experten einen Bericht über die Auswirkung großer Störfälle.

Wenn die Berichte vorliegen, werden die Regierung und dann das Parlament zum Zuge kommen. Unmittelbar nachher, das heißt gegen Ende 1983, würde gemäß heutigen Vorstellungen der Regierung eine Volksabstimmung durchgeführt. Damit wäre es nach einem Volks-Ja frühestens 1993 möglich, das erste Kernkraftwerk in Dänemark in Betrieb zu nehmen.

Bis es soweit ist, reißt die rasch wachsende Kohleeinfuhr im Vergleich zu einer entsprechenden Einfuhr von Kernbrennstoffen ein gewaltiges Loch in die Zahlungsbilanz.

Um die Größenordnungen abzustecken, sei erwähnt, daß 1980 die gesamten Einnahmen der Elektrizitätswirtschaft rund 7 Milliarden Kronen erreichten. Das sind 2% des Bruttosozialprodukts. Der Außenhandel erreichte im gleichen Jahr 109 Milliarden Kronen, der Importwert der Erdölprodukte 16 Milliarden und der Importwert der Kohle 3 Milliarden. Das heißt, die Rohenergie-Einfuhren beanspruchten ein Sechstel des Außenhandelsvolumens.

Hätte Dänemark zur Deckung der Grundlast von rund 3000 Megawatt statt Kohle- nur Kernktrftwerke eingesetzt, wäre gemäß Berechnungen aufgrund von Angaben des „Energieplans '81" unter den gegebenen Verhältnissen eine Einsparung von annähernd einer Milliarde Kronen zu realisieren gewesen. In der Tat sind auch in Dänemark die Stromerzeugungskosten seit den 50er Jahren deutlich gefallen. Aber sie liegen in den Nachbarländern Norwegen (dank der Wasserkraft) und Schweden (dank der dort bereits eingeführten Kernenergie) deutlich tiefer, und die dänischen Stromkonsumenten müssen im Vergleich zu ihren nordischen Nachbarn mit einem erheblich höheren Strompreis rechnen. Für die Industrie ist dies gewiß kein Standortvorteil.

Zusammenfassend darf gesagt werden, daß nicht nur Landschafts- und Umweltschutz für die Kernenergie sprechen. Auch handfeste betriebs- und volkswirtschaftliche Argumente können aufgeführt werden. Besonders im Hinblick auf die Versorgungssicherheit finden es die Elektrizitätsversorgungs-Unternehmen unbedingt nötig, daß Dänemark nicht ein Land ohne Kernenergie bleibt.

(SVA-Bulletin)

Mexiko. Exporterlöse in Höhe von jährlich US$ 500 Mio und ein Eigenverbrauch im Wert von US$ 300 Mio wären nach Schätzung der mexikanischen Regierung dem Land verlorengegangen, hätten die Ausbreitung der Mittelmeer-Fruchtfliege (Ceratitis) nicht aufgehalten und die bereits betroffenen Gebiete von der Plage befreit werden können. Dies gelang mit einem von der IAEA geförderten und gelenkten Programm der Sterilisation mittels Bestrahlung der männlichen Tiere.

Nachdem die Fruchtfliege in Florida in den Jahren 1929/30 mit Insektiziden erfolgreich bekämpft worden war, wurde sie in Mittelamerika im Jahre 1955 erneut beobachtet. Sie breitete sich rasch über Nicaragua und Guatemala aus und bedrohte ab 1977 den mexikanischen Gemüse- und Fruchtanbau. Die weibliche Fliege legt ihre Eier in Gemüse und Früchten ab, wo innerhalb weniger Tage die Larven ausschlüpfen, die etwa 7 bis 8 Tage von der Frucht leben und damit zu deren Fäulnis führen. Die Fruchtfliege vermehrt sich innerhalb von drei Generationen (3 Monate) auf das Tausendfache und bewirkt so, daß die Plage sich in 100 Tagen auf etwa 30000 km^2 ausbreitet.

Die konventionelle Methode der Bekämpfung der Fruchtfliege, die in tropischen und subtropischen Gebieten weltweit auftritt, erfolgt mit Insektiziden. Gegen diese Bekämpfung sprechen in Entwicklungsländern die hohen Kosten der Einfuhr der Insektizide und der flächendeckenden Anwendung. Außerdem werden Störungen des Ökosystems befürchtet. Die Sterilisationstechnik mittels radioaktiver Bestrahlung bietet demgegenüber deutliche Vorteile. Sie ist Spezies-spezifisch und umweltfreundlich und auch in wenig bevölkerten Gebieten in kontrollierbarer Weise einsetzbar.

Das von der IAEA gelenkte Programm, an dem sich die mexikanische Regierung und das US-Landwirtschaftsministerium beteiligten, wurde unter dem Namen MOSCAMED in der Zeit von 1977 bis 1981 abgewickelt. In 12tägigen Zyklen wurden jeweils ca. 2 Mia Fliegeneier gewonnen, Larven ausgebrütet und Fliegen bestrahlt. Besondere Einrichtungen zur Bestrahlung bei vermindertem Druck mußten geschaffen werden. Der Transport der sterilisierten Fliegen in die Einsatzgebiete erfolgte in gekühlten Spezialbehältern.

Als Ergebnis des Programms konnten die Nordwärtsbewegung der Fruchtfliege aufgehalten und 3 Mio Hektar Gemüseanbaugebiet durch Ausrottung der Fliege gerettet werden. Durch den weiteren gezielten Einsatz von sterilisierten Fliegen werden heute die Grenzen und die Importe kontrolliert.

(IAEA Bulletin Supplement 1982)

Schweiz. Der Bundesrat hat vom Bericht der Eidg. Kommission zur Überwachung der Radioaktivität (KUER) für das Jahr 1981 Kenntnis genommen. Die KUER, die von Prof. *Otto Huber*, Direktor des Physikinstituts der Universität Freiburg, präsidiert wird, überwacht seit 1956 die Radioaktivität der Umwelt. Die Ergebnisse des Jahresberichtes lassen sich wie folgt zusammenfassen:

Luft sowie Lebensmittel und Trinkwasser in der Schweiz wiesen 1981 nie unzulässige Aktivitätskonzentrationen auf. Alle Kernkraftwerke, das Eidg. Institut für Reaktorforschung sowie alle radioaktive Stoffe verarbeitenden Betriebe haben die Abgabevorschriften eingehalten.

Der Jahresbericht der KUER kommt zum Schluß, daß die zivilisatorische Strahlenbelastung der Schweizer Bevölkerung — mit Ausnahme des Beitrags durch medizinische Strahlenanwendungen — 1981 eine mittlere Jahresdosis von weniger als 10 mrem bewirkte. Diese ist mehr als zehnmal kleiner als die natürliche Strahlenbelastung. Die mittlere jährliche Strahlendosis durch röntgendiagnostische Untersuchungen ist demgegenüber etwa gleich groß wie die natürliche Jahresdosis.

— Weltweiter Fallout von Atombombenexplosionen: Die chinesische Atombombenexplosion vom 16. 10. 1980 bewirkte 1981 in der Schweiz eine Personendosis von ca. 0,1 mrem. Von früheren Atombombenversuchen stammendes, auf den Boden abgelagertes Caesium-137 und über die Nahrung in die Knochen eingebautes Strontium-90 ergeben immer noch Dosen von je 2 mrem pro Jahr.
— Kernanlagen und Forschungsinstitute: Die mit dem Abwasser abgegebene Radioaktivität führte bei allen Betrieben zu Jahresdosen unter 0,1 mrem, die mit der Abluft abgegebene zu solchen von höchstens 1 mrem. Die Abgabe von Jod-131 aus der Isotopenproduktion des Eidg. Instituts für Reaktorforschung in Würenlingen für medizinische Anwendungen hätte für Kleinkinder zu einer Schilddrüsendosis von 5 mrem pro Jahr (zulässig 60 mrem pro Jahr) geführt, wenn diese ausschließlich Milch von Kühen getrunken hätten, die nur am kritischen Ort grasten. Die Direktstrahlung aus den Kernanlagen führte nur unmittelbar außerhalb der Umzäunung beim Kernkraftwerk Mühleberg, beim Eidg. Institut für Reaktorforschung Würenlingen, beim Kernkraftwerk Beznau und der ehemaligen „Centrale nucléaire expérimentale de Lucens" sowie beim Schweiz. Institut für Nuklearforschung in Villigen zu erhöhten Ortsdosen. Aufgrund der kurzen Aufenthaltsdauer von Personen an diesen unbewohnten Stellen erhielt im Jahr 1981 aber niemand eine Dosis von mehr als 20 mrem. An ständig bewohnten Orten in der Umgebung dieser Anlagen lag die Ortsdosis durch Direktstrahlung unter 2 mrem pro Jahr.
— Spitäler und Industriebetriebe: In den Abwasserreinigungsanlagen Zürich und Bern wurde aus Spitälern stammendes Jod-131 festgestellt. Es verursachte im Wasser der Limmat eine mittlere Aktivitätserhöhung von 1,5 pCi/l und in der Aare eine solche von 1 pCi/l. Bei ausschließlichem Gebrauch dieses Flußwassers als Trinkwasser hätten Erwachsene in der Schilddrüse höchstens 2,5 mrem pro Jahr akkumuliert.

Wasser des Doubs unterhalb La Chaux-de-Fonds, das Tritium aus der Leuchtfarbenindustrie enthält, hätte unter gleichen Voraussetzungen zu einer Dosis von weniger als 1 mrem pro Jahr geführt.

Drei Untersuchungsreihen an Urinproben von Anwohnern der Radiumchemie A. G. Teufen/AR auf Tritium lassen den Schluß zu, daß die Tritiumabgaben dieses Betriebes über die Abluft in der unmittelbaren Umgebung zu Ganzkörperdosen von weniger als 10 mrem pro Jahr führten. In einigen 100 m Entfernung ist diese zusätzliche Dosis unmeßbar klein.

Messungen auf dem Fabrikgelände dieser Firma und auf dem ehemaligen Areal der Firma Mb-microtec A. G. in Bern-Bümpliz haben Kontaminationen des Erdbodens mit Radium und Strontium-90 ergeben. Die Beseitigung der kontaminierten Erde an diesen beiden Orten wird gegenwärtig durch die Schweizerische Unfallversicherungsanstalt (SUVA) unter Aufsicht des Bundesamtes für Gesundheitswesen veranlaßt.

(SVA-Bulletin)

Isotopenpraxis 19 (1983) 5, S. 178–184

Buchbesprechungen

Guide-lines to planning atomic spectrometric analysis Studies in analytical chemistry, Vol. 4

B. Magyar. Budapest: Akademiai Kiado 1982. 272 S., div. Abb. u. Tab.; geb. Ft 250,–

Um Elementaranalysen auf der Basis der Atomspektrometrie erfolgreich zu planen, vorzubereiten und auszuwerten, sind für den analytischen Chemiker solide Kenntnisse der physikalisch-chemischen Grundlagen dieser instrumentellen Analysenmethoden unentbehrlich. Anliegen des Buches ist es, diese Grundlagen, vor allem für den Chemiestudenten, zusammengefaßt darzustellen und dabei auf die Diskussion der Chemie der speziellen Analysen zu verzichten. Durch gute Auswahl und didaktische Aufbereitung des Materials wurde dieses Ziel erreicht.

Gegenstand des Buches sind die Methoden der Atomemissionsspektrometrie mit thermischer und elektrischer Anregung und der Atomabsorptionsspektrometrie in Flammen und Graphitöfen. Nach einem historischen Abriß und einer sachlichen Einordnung der Atomspektrometrie in die Gesamtheit der Methoden, bei denen Signale elektromagnetischer Strahlung meßtechnisch genutzt werden (Kap. 1), folgt die Diskussion der Grundelemente der Methoden. In Kap. 2 werden die wichtigsten Fakten über die Techniken der Erzeugung sichtbaren und ultravioletten Lichts sowie die Gesetze reiner Schwächung vermittelt.

Kap. 3 enthält die notwendigen Informationen über die Profile der Atomlinien in Emission und in Absorption und analysiert die Komplikationen, die sich durch den Einfluß der Selbstabsorption und die Apparatefunktionen der verwendeten Spektralgeräte auf die Abhängigkeit der Meßsignale von den Stoffkonzentrationen ergeben. Nach einem Kapitel über den Aufbau der Atomhülle sowie über Auswahlregeln, Strukturen und Intensitäten der Spektrallinien werden in Kap. 5 die Methoden der Atomisierung der Analyse beschrieben und insbesondere aus den Eigenschaften der Flamme, der Plasmafackel und der Graphitküvette die Empfindlichkeiten der Atomemissions- und Atomabsorptionsspektrometrie abgeleitet. In Kap. 6 wird auf Konstruktionsprinzipien der am häufigsten verwendeten Spektrometer und Monochromatortypen, auf die Strahlungsempfänger und die Signalverarbeitung eingegangen und werden die Grundlagen der Untergrundkorrekturen, insbesondere bei der Atomabsorptionsspektrometrie, ausführlich behandelt.

Ein mehrteiliger Anhang enthält wichtige physikalisch-chemische Angaben für den Analytiker, wie analytisch genutzte Atomübergänge und die erreichbaren Empfindlichkeiten und Nachweisgrenzen für die betreffenden Elemente. Das Buch beschreibt keine Anwendungsbeispiele und enthält nur allgemeine Hinweise zur Probenvorbereitung und -behandlung, gibt aber im Anhang eine Zusammenstellung wichtigster Monographien und Originalarbeiten, die dazu bereits zahlreich erschienen sind. Als didaktisch wertvoll erscheinen die Vergleiche der Besonderheiten und Möglichkeiten der Emissions- und der Absorptionsmethoden.

Alles in allem kann das Buch nicht nur dem noch studierenden Analytiker als Lehrmittel empfohlen werden. Es dürfte auch dem praktizierenden Analytiker von Nutzen sein, wenn er bei der Bewältigung seiner Routineanalysen sein Wissen über deren physikalische Grundlagen auffrischen möchte.

G. Müller

Hafnium: Physico-Chemical Properties of its Compounds and Alloys

Wien: IAEA 1981 (Atomic energy review, special issue No. 8) 407 S.; karton. Austr. Sch. 620,–

In der Reihe der bei der Internationalen Atomenergie-Behörde seit 1966 erschienenen Ergänzungsbände zu den Atomic Energy Reviews liegt nunmehr der 8. Teil vor. Während die ersten 6 Bände (Plutonium, Niob, Tantal, Beryllium, Thorium, Zirkon) von *O. Kubascheswki* herausgegeben wurden, zeichnet für den vorliegenden Band *K. L. Komarek* verantwortlich. Es werden wiederum die Daten über thermodynamische Eigenschaften, Phasendiagramme, Kristallstruktur, Dichte und Diffusion in einer Vollständigkeit referiert, wie sie in keiner anderen aktuellen Datensammlung zu finden sind. Der steigende Umfang der Bände (doppelte Seitenzahl im Vergleich zu den ersten drei Bänden) und die mit über 1200 Zitaten sehr große Zahl von Publikationen dokumentiert das schnelle Anwachsen des Wissensstandes über das hier behandelte Metall Hafnium sowie seine Aktualität und Bedeutung. Der Band ist wiederum gegliedert in

- Thermodynamische Eigenschaften des reinen Metalls sowie seiner Verbindungen mit Halogeniden, Bor, Kohlenstoff, Stickstoff, Sauerstoff, Schwefel und einer Reihe binärer Legierungen
  (Bearbeiter: *P. J. Spencer*. 45 S., 132 Lit.)
- Phasendiagramme mit Angaben, meist Diagrammen, zu 49 binären Systemen des Hafniums
  (Bearbeiter: *O. Kubaschewski—von Goldbeck*. 61 S., 155 Lit.)
- Kristallstruktur- und Dichte-Daten
  a) für Legierungen und Verbindungen außer Halogeniden und Chalcogeniden
  (Bearbeiter: *R. Ferro, R. Marazza*. 129 S., 645 Lit.)
  b) für Halogenide und Chalcogenide
  (Bearbeiter: *K. Girgis*. 151 S., 400 Lit.)
- Diffusion in α- und in β-Hafnium.
  (Bearbeiter: *O. Kubaschewski*. 3 S., 13 Lit.)

Der vorliegende Band ermöglicht dem Fachmann einen Überblick über den aktuellen Kenntnisstand bezüglich der physikalisch-chemischen Daten des Hafniums und erlaubt Vorausberechnungen zu solchen Fragen wie Reaktivität des Metalls, Herstellung des reinen Metalls oder seiner Verbindungen, analytische Charakterisierung. Die Datensammlung ist daher den metallurgischen Einrichtungen, in denen Hafnium oder seine Legierungen hergestellt oder mit dem Ziel spezieller Werkstoffeigenschaften bearbeitet werden, zu empfehlen. Für Einrichtungen der Reaktortechnik oder

der chemischen Industrie, in denen Hafnium oder seine Legierungen in reaktiver Umgebung eingesetzt werden, und für industrielle oder wissenschaftliche Einrichtungen, in denen Hafniumverbindungen hergestellt, eingesetzt oder untersucht werden, dürfte die umfangreiche Datensammlung ebenfalls sehr wertvoll sein.

J. Kunze

Mathematischer Einführungskurs für die Physik

S. Großmann. Stuttgart: B. G. Teubner 1981. 3. erweit. Auflage (Teubner Stud.-büch. Physik) 288 S., 92 Abb., 102 Beisp., 201 Selbsttests mit Lösungen; karton. DM 26,80

Eine wichtige Voraussetzung für die erfolgreiche Ausbildung von Physikstudenten besteht darin, daß sie ein gewisses Repertoire an mathematischen Kenntnissen und Fähigkeiten besitzen — und zwar rechtzeitig. Dieses zu vermitteln, stellt sich das vorliegende Studienbuch als Aufgabe. Es werden vor allem Studenten im ersten Studienjahr angesprochen. Daher ist das Buch auch in den ersten Kapiteln recht ausführlich gehalten und führt erst allmählich zu strafferen Formulierungen. Die Auswahl der dargelegten Gebiete (Vektoren, Vektorfunktionen, Felder, Integration, Vektorintegration, Integralsätze, krummlinige Koordinaten) erfolgte entsprechend den Anforderungen der experimentellen und theoretischen Physik. Dabei wurde auf vieles verzichtet, was die Studenten in den Grundvorlesungen Mathematik kennenlernen. Mathematische Strenge und ausführliche Beweise wurden zugunsten eines schnellen Heranführens an die mathematischen Begriffe und Methoden, wie z.B. Vektoren, Matrizen, Tensoren, grad, rot, div, Parameter-, Kurven-, Flächen-, Volumenintegrale, Integralsätze von *Gauß*, *Stokes* und *Green*, Randwertaufgaben, *Green*sche Funktionen, bewußt unterdrückt. Der Autor geht wiederholt von physikalischen Motivationen aus, um danach zu mathematisch einwandfreien Definitionen und sauberen Formulierungen von Aussagen vorzudringen.

Wünschenswert wäre es gewesen, neben dem klassischen Gebiet der Vektoranalysis auf Fragen der Wahrscheinlichkeitsrechnung und Statistik sowie auf die mathematischen Grundlagen der Quantenmechanik etwas einzugehen.

Ein wichtiges Element des vorliegenden Studienbuches besteht im Festigen des vermittelten Stoffes durch Üben. Diesem Zweck dienen die über 100 ausgearbeiteten Beispiele und die mehr als 200 Übungsaufgaben, zu denen im Anhang die Lösungen angegeben sind. Das Absolvieren dieses „Trainingsprogrammes" sei jedem Leser sehr nahegelegt. Das Verständnis vieler weiterer Betrachtungen hängt eng mit dem selbständigen und erfolgreichen Lösen der zahlreichen Aufgaben zusammen.

Die schon nach kurzer Zeit vorliegende 3. überarbeitete und erweiterte Auflage dieses mathematischen Einführungskurses zeigt, daß dieses Studienbuch einen großen Anklang gefunden hat. Und sicher umfaßt der Interessentenkreis nicht nur (wie im Titel angegeben) angehende Physiker, sondern auch zukünftige Mathematiker und Ingenieure.

G. Kummer

Differentialgleichungen

L. Collatz. Stuttgart: B. G. Teubner 1981, 6. erweit. Aufl. (Teubner Stud.-büch. Mathem.) 287 S., 173 Abb., 75 Aufgaben mit Lösungen; DM 29,80

Das vorliegende Buch stellt eine Einführung in das Gebiet der Differentialgleichungen dar. Es umfaßt sowohl klassische Teilgebiete (gewöhnliche Differentialgleichungen erster und höherer Ordnung, Systeme linearer gewöhnlicher Differentialgleichungen, Rand- und Eigenwertaufgaben, Elemente der partiellen Differentialgleichungen, numerische Methoden) als auch Teilgebiete, die in neuerer Zeit aktuell geworden sind (lineare Differentialgleichungen mit periodischen Koeffizienten, freie Randwertaufgaben, Variationsungleichungen, Methode der finiten Elemente, Verzweigungsprobleme).

Sowohl in der Auswahl des Stoffes als auch in der Art der Darstellung wendet es sich an einen Leserkreis, der an den Lösungsmethoden und Anwendungen interessiert ist. Die in vielen Lehrbüchern über Differentialgleichungen anzutreffende mathematische Strenge wird bewußt beiseite gelegt — ohne dadurch unexakt zu sein. Bei theoretischen Untersuchungen werden einfache Voraussetzungen gewählt, die bei praktischen Problemen sehr häufig erfüllt sind und die zu weniger komplizierten Beweisen führen. Mit den an technischen Hochschulen vermittelten Kenntnissen über Differential- und Integralrechnung sowie über lineare Algebra ist das vorliegende Buch voll zu verstehen. Die ersten Kapitel sind breiter und ausführlicher dargestellt, um das Verständnis für Differentialgleichungen und ihre Lösungsmethoden systematisch zu entwickeln. Durch eine große Anzahl von Beispielen und Übungsaufgaben, zu denen die Lösungen angegeben sind, wird der umfangreiche Stoff sehr anschaulich und anwendungsorientiert dargeboten. Der Leser findet rasch Zugang zu den Lösungsmethoden ohne sich durch umfangreichere theoretische Betrachtungen durcharbeiten zu müssen. Den numerischen Methoden wurde ebenfalls breiter Raum eingeräumt, da die Chance des Auffindens geschlossener Lösungen mit zunehmender Kompliziertheit der Probleme immer kleiner wird und damit die Notwendigkeit zur Anwendung numerischer Verfahren rasch zunimmt. In der vorliegenden 6. überarbeiteten und ergänzten Auflage dieses Buches wurde für die Beispiele und Aufgaben das SI-System verwendet. In Anbetracht der Gedrängtheit des Buches hätte auf die Wiederholung des Existenz- und Eindeutigkeitssatzes für gewöhnliche Differentialgleichungen verzichtet werden können.

Mehr als eine Einführung in das Gebiet der Differentialgleichungen zu geben ist bei dem beschränkten Raum nicht möglich gewesen. Daß es dem Autor gelungen ist, diese Aufgabe voll zu erfüllen, liegt an der anschaulichen Darstellung, der an den Anwendungen orientierten Auswahl des Stoffes und dem logischen Aufbau des gesamten Buches. Man kann sicher sein, daß die meisten Leser nach dem Durcharbeiten den Differentialgleichungen, denen sie in der Praxis begegnen, nicht aus dem Wege gehen, sondern sich interessiert mit ihnen auseinandersetzen werden.

G. Kummer

Sterile Insect Technique and Radiation in Insect Control Proceedings of a Symposium, Neuherberg, 27. 6.—3. 7. 1981

STI/PUB/595. Wien: IAEA 1982. 494 S., div. Abb. u. Tab.; karton.; Austr. Sch. 820,—

Die Bekämpfung von Insektenplagen durch Strahlensterilisierung (Sterile insect technique — SIT) bzw. die Anwendung von Strahlen zur genetischen Insektenbekämpfung hat in unterschiedlichen Regionen der Erde Anwendung gefunden. In dem durch die IAEA herausgegebenen Symposiumsband sind in übersichtlicher Form die letzten Entwicklungen der Anwendung der Isotopen- und Strahlentechnik zur Bekämpfung von Insekten auf medizinischem, veterinärmedizinischem und landwirtschaftlichem Gebiet dargestellt.

Zu folgenden Themenkomplexen, die weitgehend den Sektionen des Symposiums entsprechen, sind Beiträge zu finden:

- der gegenwärtige Stand der Technik zur Sterilisierung von Insekten (SIT),
- die SIT-Forschung und ihre Anwendung gegen Insekten aus medizinischer, landwirtschaftlicher und veterinärmedizinischer Sicht,
- die Tsetse-Massenanzucht,
- genetische Untersuchungen zur Strahlensterilisierung von Insekten,
- die Bekämpfung von Fruchtfliegen,
- Sterilitätsinduktion, Computersimulation,
- genetische Erkenntnisse zur Massenanzucht.

Da der Symposiumsband die einzelnen Vorträge enthält, ist das Niveau und die Form der Beiträge unterschiedlich. Allgemein handelt es sich jedoch um gut gegliederte Darstellungen mit einer Vielzahl von Literaturzitaten. Eine einheitliche englischsprachige Abfassung der Beiträge wäre wünschenswert gewesen.

K. Jung

Biologie

Ein Lehrbuch

G. Czihak, H. Langer, H. Ziegler (Hrsg.). 3., völlig neubearb. Auflage. Berlin—Heidelberg—New York: Springer 1981. 944 S., 1235 z.T. farb. Abb., 98 Tab., 2 Falttfl.; geb. DM 84,—; approx. US $ 38,20

Ein englischer Dichter des 18. Jahrhunderts — *Oliver Goldsmith* — hat sich einmal sinngemäß über gute Bücher so geäußert, daß sie ihm beim ersten Lesen gerade so erschienen, als ob er neue Freunde erwerbe. Wandelt man diesen Gedankengang auf ein gutes Lehrbuch ab, so sollte die gleiche Freundschaft, vielleicht sogar eine noch innigere, gelten und sei es nur für die Dauer des Studiums. Die 25 Verfasser des vorliegenden Lehrbuches wollen den Studenten der Biologie eine Sammlung von Grundkenntnissen anbieten, die sie während des gesamten Studiums begleitet. Die erwünschte Freundschaft zu einem guten Lehrbuch dürfte sich verstärkend auf das Eindringen und Durchdenken des dargebotenen Stoffes auswirken.

Das Lehrbuch ist in 3 große Abschnitte unterteilt: Bau und Leistung der Zelle (S. 7—166), Struktur und Funktion der Organismen (S. 169—702) und schließlich Organismen in ihrer Umwelt und in Populationen (S. 705—878) — dem letzten Abschnitt ist noch die Systematik der Pflanzen und Tiere (S. 879—902) angehängt. Es folgen Literatur- und Abkürzungsverzeichnis, das Internationale System der Einheiten (SI) und das Sachverzeichnis.

In einer 3. Auflage eines Buches sind i. allg. wenig Druckfehler zu finden, so auch in der vorliegenden Auflage; einige haben sich freilich doch eingeschlichen: auf S. 123 in Tab. 1.29 Gewebshormone nicht S. 413, sondern S. 613; auf S. 825 der Hinweis auf Abb. 6.117b S. 579 ist gemeint und nicht die Abb. 6.17b, und auf S. 860 Zwillingsarten soll der Hinweis auf S. 853 sein und nicht auf S. 85. Im Inhaltsverzeichnis wird noch Kohlendioxyd geschrieben, obwohl es dann auf den S. 531 und 537 richtig -dioxid heißt.

Folgende Verbesserungen in künftigen Auflagen könnten noch aufgenommen werden: Man sollte der Nomenklaturempfehlung der Chemiker folgen und alle sich von Äthan ableitenden Namen organischer Verbindungen mit E schreiben, beispielsweise wird dann auch die auf S. 423 für Tetraäthylammonium eingeführte Abkürzung TEA besser verständlich. Die vielen Hinweise auf Abb. sind nicht immer konsequent auch mit den entsprechenden Seitenangaben versehen, es würde eine lästige Suche nach den Abb. ersparen. Ferner wurde nicht immer die SI-Einheit für den Druck angewendet, vor allem nicht in Abb., z. B. S. 563 mm, S. 564 bar und S. 581 mbar. Die von *L. Pauling* eingeführte Masseneinheit Dalton oder Dalt sollte künftig auch vermieden werden (S. 43); sie beruht auf der nicht mehr empfohlenen chemischen Atomgewichtsskala und setzt ein hypothetisches Atom mit dem Atomgewicht 1 voraus. Die zeitliche Änderung der Populationsgröße N ist nicht exponentiell, sondern quadratisch von N abhängig (S. 742 — Text und Legende zu Abb. 8.49), wie überhaupt die gesamte Argumentation auf der Diff.-Gleichung 8. 10 *c* S. 738 beruht und nicht auf deren Lösung $N = K \cdot N_0 [N_0 + (K - N_0) \exp(-r_0 t)]^{-1}$. Auf S. 730 sollte eine Umrechnungsbeziehung der verschiedenen möglichen Angaben für die Biomasse eingefügt werden. Auf S. 771 wird gesagt, daß derzeitig die Kernenergie für die Energieerzeugung noch vernachlässigt werden könnte — wohl kaum: Im Jahre 1981 betrug der Anteil der Kernenergie an der Weltelektrizitätserzeugung ungef. 9% entsprechend 952603 MW(e), die in 272 Leistungsreaktoren erzeugt wurden. Man schätzt, daß dieser Anteil 1985 auf ungefähr 17% ansteigen wird. Welche Folgen sich daraus auch für die Biosphäre ergeben, wird überhaupt nicht berührt (langlebige Radionuklide in Nahrungsketten, Radioökologie u. a. m.). Die Anwendung von Isotopenmethoden in der Biologie wird auf den S. 102, 104, 172, 213, 295, 311, 315, 590 und 633 erwähnt, nicht aber im Inhaltsverzeichnis. Das Alter für die Geschichte (Evolution) des Lebens wird an 4 Stellen (S. 37, 784, 789 und 818) erwähnt, allerdings mit 3 unterschiedlichen Angaben ($4 \cdot 10^9$ a, $>3 \cdot 10^9$ a und zweimal $>2 \cdot 10^9$ a). Hier sollte nicht nur eine einheitliche Auffassung vertreten werden, sondern auch an die Isua-Sedimente von Westgrönland gedacht werden, die mit $3{,}8 \cdot 10^9$ a datiert wurden und auf erste Lebensspuren hindeuten. Es sollten überhaupt die geologischen Einordnungen spätes Tertiär, Miozän, Pleistozän u. ä. durch entsprechende Altersangaben ergänzt werden, um sich ein besseres Bild von den abgelaufenen Zeiträumen machen zu können. Als die Physiker des 19. Jahrhunderts für die Strahlungsenergie der Sonne nur die Gravitationsenergie kannten und eine darauf beruhende Berechnung des Alters der Sonne nur eine Zeitspanne von $30 \cdot 10^6$ a ergab, waren es schließlich die Biologen, die auf diesen Widerspruch hinwiesen und für die Evolution eine wesentlich längere Zeitspanne als notwendig ansahen. Auf S. 868 wird die „Eroberung" des Landes durch die Pflanzen nur als ein Wechsel im Umweltbezug erklärt. Es sollte doch gesagt werden, daß der durch die Photosynthese gebildete Sauerstoff eine Konzentration in der Atmosphäre erreichte, die den schützenden Ozonschild ausbilden konnte.

In einer Zeit, in der die moderne Biologie nicht nur ein philosophisch-weltanschauliches Gebäude darstellt, sondern sogar als politisch-ideologisches Programm dienen und ausgebaut werden kann, sollten einige Aspekte in dieser Richtung Erwähnung finden (Ökologie und Ökonomie, Monokulturen, Landschaftsschutz und Bodenerosion, Tier- und Pflanzenschutz, Walfang, Pflanzenschutzmittel, Spurenmetalle in Nahrungsketten u. a. m.). Die Literatur ist mit insgesamt 266 Hinweisen erfaßt worden, wovon rund 66% aus den Jahren 1971—81 stammen — eine aktuelle Zitation. Manche der schönen Abbildungen sollten nicht nur mit dem Autorennamen belegt werden, sondern möglichst mit der Literaturstelle. Im Vorwort zur 1. Auflage dieses Lehrbuchs heißt es u. a., daß die Autoren mit Begeisterung an diesem Buch gearbeitet haben. Diese Begeisterung kommt an vielen Stellen in der Darstellung des umfangreichen Stoffes zum Ausdruck. Das Buch verspricht in seiner neuen Art, die Freundschaft der Leser und Studenten zu erringen — es ist ein sehr gutes Lehrbuch über eine schöne Wissensdisziplin, die Lehre vom Leben.

H. Hübner

Erlebte Wissenschaft
Berühmten Forschern über die Schulter geschaut

J. Bernstein. Düsseldorf: Econ Verl. 1982. 304 S.; geb. DM 42,—

Der Autor, selbst Theoretiker auf dem Gebiet der Elementarteilchenphysik und Lehrstuhlinhaber für Physik am Stevens Institute of Technology, New York, betrachtet sich selbst als Wissenschaftskritiker. Er hat mehrere Beiträge im „New York Times Magazine" und im „New Yorker" dazu verfaßt und legt in dem Buch einen Teil dieser Beiträge vor. Außer einer Einführung, einigen Anmerkungen des Übersetzers besteht das Buch aus 3 großen Teilen: einem Teil, der sich mit Physik befaßt, einem über Biologie und einem über Rechentechnik, mathematische Logik bzw. Erkenntnistheorie, vereinfacht gesagt.

Die Physik wird in zwei Ansichten dargestellt (S. 14—149): die Welt *Johannes Keplers* und die Welt des jüngst 84 Jahre alt gewordenen Nobelpreisträgers *Isidor Isaac Rabi*. Für beide wird versucht, ihr Ringen um Erkenntnisse und ihre Motivationen darzustellen, wobei der Autor bei *I. I. Rabi* Gelegenheit hatte, durch Interviews manches nachzufragen und nachträglich aufzuklären. Welchen zähen Kampf *J. Kepler* mit sich führen mußte, ist den heutigen kurzen Darstellungen seiner bekannten Gesetze in Lehrbüchern nicht mehr anzusehen.

Die Biologie wird in drei Ansichten dargestellt (S. 152 bis 225): *Trofim Denissowitsch Lyssenko* (1898—1976), *Rosalind Franklin* (1920—1958) und *Lewis Thomas* (1913 — seit 1973 Professor für Medizin und Pathologie an der Cornell-Universität in Ithaca im Staate New York); *T. D. Lyssenko* und sein Wirken werden auf 10 Seiten dargestellt. Als einzige und angeblich beste Quelle wird ein Buch eines im Ausland lebenden Dissidenten verwendet, während die Arbeiten sowjetischer Genetiker, wie die von *Sergej Sergejewitsch Tschetwerikow* (1880—1959), von *Nikolaj Petrowitsch Dubinin* (1907) u. a. gar nicht erst erwähnt werden, auch nicht die langjährigen Kontroversen von *Lyssenko* gemeinsam mit *Gluschtschenko* mit den genannten Genetikern. Hier scheint der Autor den Forschern nicht nur über die Schulter geschaut zu haben, sondern hier ist leicht eine Absicht erkennbar. Das Wirken von *Rosalind Franklin* wird da schon kritischer und tiefer betrachtet, denn sie stand mehrmals vor der Aufklärung der Struktur des DNS-Moleküls und hatte doch niemanden in ihrer Nähe, mit dem sie sich hätte darüber aussprechen können — ein Beispiel dafür, wie notwendig und wirkungsvoll eine sich gegenseitig befruchtende Zusammenarbeit von Wissenschaftlern sein kann (siehe *Watson* und *Crick*).

Die letzten drei Essays sind mit „Faktisches und Fantastisches" überschrieben und berühren Fragen der Art, ob neue und zwar wirklich neue Ideen auch gleichzeitig verrückt sein müssen. *Arthur C. Clarke* zeichnete sich durch genug verrückte Ideen aus; beispielsweise hatte er im Oktober 1945 die Idee der Nachrichtenübertragung durch Satelliten, zu einer Zeit also, in der die Raumfahrt aus heutiger Sicht noch unbekannt war. Die Frage nach der Vollständigkeit von Axiomensystemen, die u. a. der österreichische Mathematiker *Kurt Gödel* 1931 betrachtet hat, ist wohl ohne nähere Einzelheiten in Ausführung und Beweisführung nicht darstellbar. Wenn man beweisen kann, daß *a* und gleichzeitig nicht *a* gilt, dann liegt offensichtlich ein Widerspruch vor. Die Axiomensysteme der Arithmetik, der Geometrie und der Mengenlehre bedürfen aber eines weiteren Axioms, daß eine der beiden Aussagen bewiesen werden kann, um ein vollständiges Axiomensystem zu sein, was *Gödel* zeigen konnte und was heute noch eine fast ans Unglaubliche reichende Feststellung ist. Welches Erwachen für einen Mathematiker, bewiesen zu bekommen, daß seine wichtigsten Axiomensysteme unvollständig sind, jedenfalls kategorisch unvollständig sind und daß sie eine Ergänzung der Wahl bedürfen, welche der beiden Aussagen gelten soll, um in sich widerspruchsfrei zu sein. Hier beschreibt der Autor mehr sein eigenes Ringen um das Verstehen des *Gödel*schen Theorems als das Ringen *Gödels*.

Das Buch ist an all den Stellen interessant und wissenserweiternd, wo der Autor durch unmittelbare Gespräche die von ihm charakterisierten Wissenschaftler vorstellt, oder wo er ein umfangreiches Quellenstudium vorgenommen hat (Beispiel *Kepler*), an anderen Stellen bedarf es noch mehrfacher Blicke über die Schulter, um die Wissenschaft voll zu erleben, es sei denn, man liest die Beiträge als ein Feature, ohne die Wissenschaft unmittelbar erleben zu wollen.

Bleibt zu fragen, was ein Wissenschaftskritiker ist: Vom Kritiker stammt offensichtlich der Teil, der subjektiv die Einschätzung oder Beurteilung von Werken oder Taten vornimmt, von der Wissenschaft das Objektive, das sich aus den Ansichten der Zeit heraus nicht wandelt. Vielleicht genügt es auch, nur zu behaupten, ein (Wissenschafts)-Kritiker zu sein oder doch nur ein Kritiker der Wissenschaftler? Das Buch jedenfalls kann diese Fragen nicht klären.

H. Hübner

Die Geschichte der Radioaktivität

C. Keller. Stuttgart: Wissenschaftliche Verlagsgesellschaft 1982; 110 S., 39 Abb., 8 Tab. (Bücher der Zeitschrift Naturwissenschaftliche Rundschau) Kst. geb. DM 32,—

Die faszinierende Geschichte der Entdeckung der natürlichen Radioaktivität, der künstlichen Radioaktivität und der Kernspaltung wird immer wieder in der Literatur, wenn auch unter unterschiedlichen Aspekten, dargestellt. Das liegt nicht zuletzt daran, daß es sich hier im wahrsten Sinne des Wortes um eine „Bilderbuchgeschichte" handelt, in der alle Elemente erfolgreichen Forschertums, wie Gewinnung grundlegender neuer Erkenntnisse der Naturwissenschaft bei geringem materiellen Aufwand, kaum zu übersehende Konsequenzen für Wissenschaft und Technik, ja für die gesamte Menschheit, zum Ausdruck kommen. Das alles wurde von wenigen genialen Menschen, die sozusagen nur in die Kiste greifen, aber richtig deuten mußten, absolviert. Dabei spielt der Zufall, das unerwartete wissenschaftliche Ergebnis und dessen konsequente Verfolgung eine überragende Rolle. Selbst Elemente des wissenschaftlichen Meinungsstreits, Prioritätsansprüche, Werturteile über Fachkollegen, bis hin zu kleinsten privaten Reminiszenzen und Anekdoten, fehlen nicht. Schließlich wird gezeigt, daß festgelegte Meinungen, Theorien, besonders von den „Päpsten", auch den Fortschritt hindern bzw. verlangsamen können. Diese interessanten Dinge können, auch gegenwärtig noch, von solchen Wissenschaftlern, die zumindest eine längere

Wegstrecke dabei waren, sozusagen aus erster Hand dargestellt werden. Das alles ist Wissenschaftsgeschichte über nur wenige Jahrzehnte, die noch durch zwei verheerende Weltkriege unterbrochen und in denen in spektakulärer Weise Erkenntnisse friedlichen Forschertums zur Massenvernichtung von Menschen eingesetzt wurden; welch ein Stoff für unterschiedlichste Denkansätze.

Das *Keller*sche Buch ist neben Vorwort, Anhang und Verzeichnissen in die folgenden 5 Abschnitte eingeteilt:

Prolog, Entdeckung der natürlichen Radioaktivität	(15 S.)
Der Weg zur künstlichen Radioaktivität	(4 S.)
Die falschen Transurane	(9 S.)
Die Entdeckungsgeschichte der Transurane	(26 S.)
Transurane in der Natur	(33 S.)

Wie das bereits im Titel zum Ausdruck kommt, ist hier besonders die Geschichte der Transurane (etwa $^2/_3$ des Buches) dargestellt. Dadurch unterscheidet sich dieses Buch von der Vielzahl anderer geschichtlicher Abhandlungen. Ein weiterer Vorteil dieses Werkchens liegt darin, daß sich der Autor doch relativ streng an die wissenschaftlichen Zusammenhänge hält und auf Späßchen, Reminiszenzen und umfangreiche Allchemie-Betrachtungen weitgehend verzichtet. Dadurch wird der Leser nicht nur unterhalten, sondern dem Lernenden bzw. weniger Eingeweihten werden durchaus Informationen und exaktes Faktenwissen vermittelt.

Man kann sicher unterschiedlicher Meinung darüber sein, ob zum Kapitel „Transurane in der Natur", wie das hier geschehen ist, auch der Anteil an Transuranen gehört, der durch den Menschen versehentlich oder auch bewußt (fall out, Atombombenflugzeug-Katastrophen, unsachgemäße Behandlung etc.) in die Umwelt gelangt ist.

In einer Geschichte der Radioaktivität sind natürlich auch Fragen der Erzeugung superschwerer Elemente, Kriterien für die Entdeckung neuer Elemente bzw. deren einwandfreie Identifizierung doch wohl mehr Zukunftsaspekte. Insgesamt gesehen: eine didaktisch sehr gut aufgebaute Schrift — das erwartet man allerdings vom Autor —, die sicher von historisch interessierten Naturwissenschaftlern, wie Radiochemikern, Kernphysikern usw., wie auch von naturwissenschaftlich interessierten Laien, gern gelesen wird.

H. Koch

High-Energy Ion-Atom Collisions

Proc. of the Internat. Seminar on High-Energy Ion-Atom Collision Processes, Debrecen 17—19, March 1981

D. Berényi, G. Hock (Hrsg.). *Budapest: Akadémiai Kiado 1982: 275 S., zahlr. Abb. u. Tab. geb. M 97,50*

Im März 1981 veranstaltete das bei seinen internationalen Kontakten überaus rührige Institut für Kernphysik der Ungarischen Akademie der Wissenschaften (Debrecen) unter der Schirmherrschaft des Nestors der ungarischen Kernphysik, *A. Szalay*, ein internationales Seminar zu dem mit dem Buchtitel identischen Thema, das auf dem Wege dieses Instituts insofern einen Meilenstein darstellt, als dieses Seminar den Einstieg in ein neues, längerfristiges Arbeitsgebiet des Instituts markiert. So steht denn auch das Referat des Institutsdirektors Prof. *Berenyi* „A research program for ion-atom collisions at high energy" im Mittelpunkt des Ganzen. Von den ungefähr 100 Fachvertretern bzw. Autoren des Seminars aus fast allen Kontinenten mit in der Fachwelt teilweise sehr klangvollen Namen erwarteten sich die ungarischen Kollegen helfende Kritik für ihr zukünftiges Programm und darüber hinaus manche weiterführende Anregung. Aber selbstverständlich trifft zumindest der Gesichtspunkt der wissenschaftlichen Anregung wechselseitig und ganz allgemein auf alle Teilnehmer dieser von der Atmosphäre her relativ intimen Veranstaltung zu.

Methodisch liegt der Schwerpunkt auf Seiten der Elektronenspektroskopie, d. h., es ist sehr viel seltener die Rede von Röntgeneffekten. Hierfür sind die Beiträge von *L. H. Toburen*, *D. Schneider* und *N. Stolterfoht*, *R. Mann* und *J. P. Briand* einige Beispiele aus der Reihe der vollständig abgedruckten Plenar- bzw. Hauptvorträge.

Von den Kurzvorträgen des Seminars findet sich in dem Band nur eine kleine Auswahl wieder, denn meist ist ja ihr Inhalt Gegenstand von Originalmitteilungen in Fachzeitschriften.

Die sehr spezielle Thematik und wohl nicht zuletzt der hohe Preis des Buches werden den Leserkreis sicherlich in engen Grenzen halten: Es ist eine Publikation für Bibliotheken von Spezialinstituten für Atom- und Kernphysik.

G. Brunner

Finite Modelle gewöhnlicher Randwertaufgaben

E. Bohl. Stuttgart; B. G. Teubner 1981; 318 S., 21 Abb., 76 Aufgaben, zahlr. Beispiele (Leitfäden der angewandten Mathematik und Mechanik Bd. 51 — Teubner Studienbücher) kart. DM 29,80

„Dieses Buch gibt eine Einführung in die Konstruktion finiter Modelle zur numerischen Behandlung einer Randwertaufgabe

$$\text{(D)}\quad (-px')' + \nu(kx)' = \mathrm{f}(t, x, \lambda) \quad \text{in} \quad [\mathrm{a}, \mathrm{b}]$$

$$\text{(R)}\quad R_{\mathrm{a}}x = \alpha_{\mathrm{a}}x(a) - \beta_{\mathrm{a}}x'(a) = \gamma_{\mathrm{a}}$$

$$R_{\mathrm{b}}x = \alpha_{\mathrm{b}}x(b) + \beta_{\mathrm{b}}x'(b) = \gamma_{\mathrm{b}}$$

Es geht um Differenzenverfahren und Verfahren finiter Elemente, welche auf lineare und nichtlineare Gleichungssysteme besonderer Struktur führen. ... Gleichzeitig wollen wir in die Anwendungen einführen. Es werden Beispiele aus der Physik, der Chemie und der Biologie zusammengetragen und deren Herkunft beleuchtet."

Das vom Autor in seiner Einleitung so umrissene Programm wird in zehn Kapiteln entwickelt. Ausgehend von typischen Beispielen aus Chemie und Physik wird zunächst relativ elementar das klassische Differenzverfahren für die parameterfreie lineare und nichtlineare Gleichung (D) diskutiert, um genügend Motivation für einen Abriß der Theorie von Iterationsprozessen zur Lösung nichtlinearer Gleichungssysteme (Kap. III) und eine allgemeine Konvergenztheorie für finite Modelle (Kap. IV) zu haben. Anhand der spezielleren Gleichungen

$$-x'' = \mathrm{f}(t, x) \quad \text{und} \quad -x'' + k(t)\, x' = \mathrm{f}(t, x)$$

werden dann Differenzenverfahren höherer Ordnung diskutiert und die Methode von *Galerkin* zum Ausgangspunkt für ein Verfahren der finiten Elemente behandelt. Beispiele der Reaktionskinetik führen dann in den Kapiteln VII—IX auf der Grundlage der behandelten numerischen Verfahren zu einem experimentell-numerisch gestützten Verständnis wichtiger Phänomene, wie der Lösungsverzweigung, der Hysteresis oder singulärer Störungen bei der Ausbildung von Grenzschichten. Ein Abriß der mathematischen Lösungstheorie der Zwei-Punkt-Randwertaufgaben stellt schließlich eine Reihe früher gewonnener Aussagen auf ein sicheres Fundament und rundet das Buch zusammen mit den jedem Kapitel beigegebenen Literaturhinweisen (ein Verzeichnis von etwa 140 meist jüngeren Arbeiten findet sich am Ende des Buches) in schöner Weise ab.

Man merkt es dem Buch an, daß es die Frucht intensiver Lehrtätigkeit ist, und der Rezensent hält es für den akademischen Lehrer, welcher Studenten der Naturwissenschaften die Anwendung mathematischer — hier vor allem numerischer — Methoden auf ihr Fachgebiet nahebringen will, und für die betroffenen Studenten selbst für außerordentlich wertvoll, insbesondere dann, wenn direkt an einer dialogfähigen Rechenanlage experimentiert werden kann, was der Autor besonders empfiehlt.

Die diskutierten Probleme sind eben tatsächlich aus den naturwissenschaftlichen Theorien hergeleitet und stellen nicht nur naturwissenschaftlich formulierte akademische Beispiele dar.

Das Bestreben, auf möglichst engem Raum sehr viele Gedanken, Fakten und schließlich auch noch didaktische Absichten unterzubringen, kann allerdings die Lesbarkeit an manchen Stellen spürbar beeinträchtigen. So werden etwa am Anfang ohne eigentliche Motivation eine ganze Reihe von Spezialbezeichnungen und Verweiskürzeln eingeführt, und später einfach benutzt. Der Leser neigt aber wohl dazu, erst dann Bedürfnis nach kurzen Bezeichnungen zu entwickeln, wenn er sachlich dazu motiviert wird. So kommt er in die Lage, ständig zurückblättern zu müssen und wird im Gedankenfluß nachhaltig gehemmt.

Typographie und drucktechnische Ausführung sind hervorragend, und die Zahl der Druckfehler hält sich in sehr engen Grenzen.

So kann das Buch allen denen empfohlen werden, die ein echtes Arbeitsbuch über den Gegenstand suchen: dem eingeweihten Hochschullehrer vor allem als willkommene Bereicherung seines Schatzes instruktiver (und durchgerechneter!) Beispiele, dem Studenten als wertvolle Ergänzung zu einer entsprechenden Lehrveranstaltung und dem Naturwissenschaftler, der sich mit der numerischen Durchdringung seiner Wissenschaften auseinandersetzen und sich nicht damit begnügen möchte, Ergebnisse von einem Rechenzentrum kommentarlos übernehmen zu müssen.

R. Hofmann

Modern plasma physics

Trieste Course 1979

Vienna: IAEA 1981; 608 S., div. Abb. u. Tab. (STI/PUB/573) kart. ÖS 930,—

Dieses Buch berichtet von einem Kurs über Methoden und Ergebnisse der modernen Plasmaphysik am Internationalen Zentrum für theoretische Physik in Triest vom 16. Oktober bis 23. November 1979 und enthält im Teil I sieben Vorträge des Grundkurses (S. 3—304). Der Teil II enthält acht ausgewählte Vorträge über Probleme der Fusionsforschung (S. 307—490). Abschließend werden im Teil III zwei Übersichtsartikel (S. 491—592) abgedruckt.

Der Anhang enthält das Programm der Schule, die Namen der Mitglieder des Wissenschaftlichen Rates und die Teilnehmerliste.

Im Teil I findet man Vorträge über Plasmaeigenschaften (*P. H. Sakanaka*), Fragen der Stabilität und Instabilität, insbesondere in fluiden Plasmen und Fusionsanlagen (*S. Cuperman, A. A. Skorupski, A. M. El Nadi*). Weiter werden nichtlineare Effekte bei der Wellenfortpflanzung behandelt. Eine wichtige Rolle in diesem Teil spielt die Behandlung kinetischer Gleichungen wie *Boltzmann-*, *Liouville-*, *Fokker-Planck-*, *Vlasov-* und *Balescu-Lenard*-Gleichung (*M. H. A. Hassan*).

Der in Teil I dargestellte Grundkurs gibt einen umfassenden Überblick über eine Vielzahl von Problemen der modernen Plasmaphysik. Teil II wendet sich aktuellen Fragen der Fusionsforschung zu. Hier wären aus der Sicht des Rezensenten besonders die Vorträge von *Engelmann* (Transport in toroidalen Magnetoplasmen), von *J. Killeen, A. A. Mirin, M. G. McCoy* über ein Transportmodell für Tokamaks und der Vortrag von *A. A. Offenberger* über Experimente mit hochintensiven CO_2-Lasern hervorzuheben.

Der Teil III enthält schließlich die Vorträge von *B. Lehnert* „Small-scale Experimente in poloidalen Feldsystemen" und *A. A. Rukhadze* „Wechselwirkung von relativistischen Elektronenstrahlen mit Plasma", in dem die physikalische Natur der Phänomene und die Anwendung in der Mikrowellenelektronik untersucht werden.

Die vorliegende Ausgabe vermittelt dem Leser in gedrängter Form einen sehr guten Überblick über eine große Palette aktueller Fragen.

R. Haberlandt

Im Schatten der Sensation

Leben und Wirken von Fritz Straßmann

F. Krafft. Weinheim: Verlag Chemie 1981; 541 S., 48 Abb., zahlr. Tab. u. Faksimiles; Leinen DM 150,—

Mit Ehrungen und Auszeichnungen ist es so eine Sache: Wer von den unmittelbar Beteiligten ist objektiv genug, eine wirklich sachliche Einschätzung zu liefern, und wer von einer „Jury" hat tatsächlich hinreichenden Einblick? Aber es ist ja meist noch viel schlimmer. Selbst besten Willen vorausgesetzt: Was darf denn in solchen Zusammenhängen objektiv oder wirklich sachlich genannt werden? Für das schwedische Nobelkomitee jedenfalls gehörte mindestens auch noch Mut dazu, nicht viel länger als ein Jahr nach dem Abwurf der beiden Atombomben über japanischen Großstädten, sozusagen in *Lise Meitners* schwedischer Nachbarschaft dem Deutschen *Otto Hahn* den Nobelpreis für Chemie des Jahres 1944 zuzuerkennen.

Um die Zeit des 60. Geburtstages *Otto Hahns* (8. 3. 39) gab es wohl einiges Gerangel um die Priorität der Entdekkung der Uranspaltung, und vielleicht läßt sich Anliegen wie Inhalt des Buches am besten und subjektiv am unverdächtigsten (für den Buchautor wie für den Rezensenten!) durch ein Zitat aus einem Brief *O. Hahns* an *L. Meitner* vom 3. 3. 39 — also aus diesen streitbaren Tagen — kennzeichnen: „Der Einzige, der diese Dinge [d. h. die Prioritätsfragen, G. B.] absolut ruhig beurteilt, ist *Straßmann* selbst, der — das muß man sagen — in jeder Beziehung am schlechtesten wegkommt." (Zitat aus *F. Krafft*, S. 325).

Als dann Jahrzehnte später, nämlich am Tage nach *F. Straßmanns* 60. Geburtstag (22. 2. 62) im Berliner „Tagesspiegel" eine von *Alexander Marfeld* verfaßte Würdigung des Jubilars erschien, trug diese die bezeichnende Überschrift „Im Schatten der Sensation. Zum 60. Geburtstag von Professor Dr. Fritz Straßmann".

Prof. *F. Krafft* als Buchautor ist nicht Kernchemiker oder -physiker, sondern Wissenschaftshistoriker. Sein Buch über *F. Straßmann* ist Ausdruck einer sehr hochrangigen wissenschaftlichen Leistung. Sie erforderte tiefes Eingehen auch auf chemisch-physikalische Details. Was *Krafft* hier zusammengetragen hat, ist überzeugend und beeindruckend. Das gilt nicht nur in bezug auf den im Rahmen seiner Zeit porträtierten stillen, überaus bescheidenen aber excellenten radiochemischen Analytiker *F. Straßmann*, sondern auch für den Buchautor selbst. Dabei war (und ist) es offenbar nicht in jedem Falle leicht, an das wissenschaftshistorische Material heranzukommen; und auch an Polemik von dritter Seite im Zusammenhang mit der Akzentuierung von *Straßmanns* Anteil an der Arbeit des Teams mit *Hahn* und *Meitner* hat es offenbar nicht gefehlt, wenn man z.B. die Reaktionen von *Otto Hahns* Enkel, *Dietrich Hahn*, auf schon etwas zurückliegende Vorträge und Publikationen *F. Kraffts* zum gleichen Thema zur Kenntnis nimmt (siehe z.B. Phys. Blätter **37** (1981) 44–46).

Es gibt viel zu lernen beim Studium dieses – man möchte schon sagen – einmaligen Werkes: auch oder gerade weil die Gedanken an Sinn und Zweck der Forschung, die sich auf dem unterlegten und emotional so belasteten Fachgebiet so oft nahelegen, eben nicht zwangslos in landläufige Klischees einmünden; aber auch sonst: Wie berührt heute besonders junge Menschen die Tatsache, daß der 32jährige *Straßmann* einen 600,– M-Posten als Chemiker bei der Solvay-AG ausschlug und sich statt dessen, am Kaiser-Wilhelm-Institut verbleibend, aber „freiberuflich" tätig, mit weniger als einem Viertel dieses Geldes durch Nachhilfeunterricht über Wasser hielt? Dabei ging es offenbar weniger um den Idealismus der „hehren Wissenschaft" gegenüber als vielmehr darum, aus Abneigung gegen das herrschende Regime nicht einer offiziellen Organisation (Bedingung bei Solvay!) beitreten zu müssen. Ähnliches war übrigens auch der Grund für das Nichtvollziehen der Habilitation.

Das Buch *F. Kraffts* ist eine wissenschaftsgeschichtliche Fundgrube par excellence und vermittelt wissenschaftsethische und menschliche Haltungen des Teams *Straßmann, Meitner* und *Hahn*, die jeder Naturwissenschaftler nur mit Hochachtung zur Kenntnis nehmen kann.

G. Brunner

Terrestrial Nitrogen Cycles

Processes, Ecosystem Strategies and Management Impacts

F. E. Clark, T. Rosswall (Ed.). Stockholm: Swedish Natural Science Research Council 1981; 714 S., 169 Abb., 121 Tab. (Ecological Bull. No. 33) geb. DM 116,–

Das Studium von biogeochemischen Kreisläufen bedarf der internationalen Zusammenarbeit von Wissenschaftlern in einem besonders hohen Maße. Deshalb initiierte auch das Scientific Committee on Problems of the Environment (SCOPE) des Internationalen Rates der Wissenschaftsvereinigungen (ICSU) bereits 1974 internationale Gremien, auf welchen einzelne oder mehrere biochemisch relevante Elemente bearbeitet und diskutiert wurden. Bekannt geworden sind die Ergebnisse durch die SCOPE Reports, beispielsweise der Report 7 über die Elemente Stickstoff, Phosphor und Schwefel (1976) oder der Report 14 über den globalen Kreislauf des Kohlenstoffs (1979).

Vom 16.–22. September 1979 fand in Gysinge Värdshus, Österfärnebo (Schweden) eine internationale Arbeitstagung statt, die sich nur mit dem Kreislauf des Stickstoffs befaßte und auf der 38 Vorträge über dieses Thema gehalten wurden. Die Tagung wurde u. a. von der Schwedischen Akademie der Wissenschaften organisiert. An der Tagung nahmen 66 Wissenschaftler aus verschiedenen Ländern teil, außerdem 6 Beobachter internationaler Wissenschaftsagenturen (UNEP; IAEA/FAO) oder nationaler Wissenschaftsorganisationen (USNSF; The Royal Swedish Academy of Agriculture and Forestry u. a.). Der vorliegende Band enthält die auf dieser Tagung gehaltenen Vorträge, einschließlich der Ergänzungen durch Berichterstatter (Raporteurs), einer Schlußfolgerung für künftige Arbeiten zum terrestrischen N-Kreislauf und einem Inhaltsverzeichnis, was i. allg. bei Proceedings nicht vorgefunden wird, aber sehr zweckmäßig ist.

Von den bedeutsamen Prozessen, die am N-Kreislauf teilnehmen, wurden sowohl die biologischen (N-Aufnahme, Nitrifikation, Denitrifikation, N_2-Fixierung u. a.) als auch die teilweise schwierigeren physikalisch-chemischen Vorgänge (trockene und nasse Ablagerung von N-Verbindungen, Mobilität des NH_4^+ im Boden u. a.) getrennt und ausführlich betrachtet (S. 123–374). Besonders hervorzuheben sind aber die Ausführungen, die einzelne Ökosysteme als geschlossene Einheiten ansehen und die darin ablaufenden N-Kreisläufe analysieren (S. 375–528). So werden die Tundren, die Savannen, die Graslandschaften und Waldzonen zur Darstellung gebracht und ihre Besonderheiten hervorgehoben. Wenn auch diese Subökosysteme nur Teile des globalen N-Kreislaufes sind und eine geschlossene Bilanzierung ohne die aquatischen Ökosysteme unserer Erde gar nicht möglich ist, so bilden diese Subsysteme doch wesentliche Einflußfaktoren auf das globale Geschehen und zeichnen sich auch teilweise durch eine komplexe Verflechtung sehr verschiedener Vorgänge aus.

Was immer besonders wertvoll bei Proceedings ist und so auch hier hervorgehoben werden kann, ist die Aktualität der Literaturzitate. Es werden insgesamt 2337 Arbeiten zitiert, wovon 70,7% aus den Jahren 1971–1981 stammen. Das Maximum der Zitate liegt bei 1978 mit 12,6%; alles in allem eine Fundgrube nicht nur an dargebotenem Wissensstoff, sondern auch an aufgeführter und weiterführender Literatur. Ferner ist auffällig, wie zwanglos und ohne erkennbare Betonung die Anwendung von Isotopenmethoden beim Studium der Teilprozesse des N-Kreislaufes einbezogen wurde. In 8 Arbeiten werden ^{15}N-Markierungen angewendet, in 3 Arbeiten ^{14}C-Markierungen. Keine Arbeit befaßte sich mit dem Studium natürlicher Isotopenvariationen des Stickstoffs in der Bodenzone bzw. in der terrestrischen Biosphäre, ein Gebiet, das gerade in den letzten Jahren bemerkenswerte Beiträge zum Verständnis der Prozesse lieferte.

Daß der vorliegende Proceedings-Band für Ökologen eine wertvolle Datensammlung darstellt, ist vordergründig leicht erkennbar. Das Buch dürfte aber auch Biochemikern, Biophysikern und Agrar- bzw. Forstwissenschaftlern eine Quelle für Einblicke in die Vielfalt der ablaufenden Prozesse sein.

H. Hübner

Contents of this Issue

Articles

Letters to the Editor

Indexed in Current Contents/PC & ES
Science Citation Index
and INIS Atomindex